BUILDING
THE UNIVERSE

NEW SCIENTIST GUIDES
Series Editor: Michael Kenward

BUILDING
THE UNIVERSE
Edited by
CHRISTINE SUTTON

Basil Blackwell and New Scientist

First published in book form in 1985 by
Basil Blackwell Limited.
108 Cowley Road, Oxford OX4 1JF.

Basil Blackwell Inc.,
432 Park Avenue South, Suite 1505,
New York, NY 10016, USA.

British Library Cataloguing in Publication Data

Building the universe.——(New Scientist guides)
  1. Nuclear reactions   2. Particles (Nuclear physics)
  I. Sutton, Christine   II. Series
  539.7'54   QC794

    ISBN 0-631-14102-2
3 7 6 ⁹²¹² 0-631-14103-0 Pbk

Library of Congress Cataloging in Publication Data
Main entry under title:

Building the universe.
(A New Scientist guide)
Includes index.
  1. Nuclear physics   2. Particles (Nuclear physics)
  I. Sutton, Christine   II. Series
  QC776.B83 1984   539.7   84-14454

  ISBN 0-631-14102-2
  ISBN 0-631-14103-0 Pbk

Typeset by Katerprint Co Ltd, Oxford
Printed and bound in Great Britain by
T. J. Press, Padstow

Contents

Contents

Contributors

ROGER BLIN-STOYLE is professor of theoretical physics at the University of Sussex.

FRED BULLOCK is in the Department of Physics and Astronomy at University College, London. He was a member of the team that found the first evidence for neutral currents at CERN.

FRANK CLOSE is in the theory division of the SERC's Rutherford Appleton Laboratory, Oxfordshire. He is author of *The Cosmic Onion: Quarks and the Nature of the Universe* (Heinemann Educational, 1983).

PAUL DAVIES is now professor of theoretical physics at the University of Newcastle upon Tyne. He is author of many popular books on physics, his latest being *God and the New Physics* (Dent, 1983).

JAMES DODD writes about particle physics and is author of *The Ideas of Particle Physics* (CUP, 1984).

NORMAN DOMBEY is a theoretical physicist in the School of Mathematical and Physical Sciences at the University of Sussex.

JOHN ELLIS is currently working on theories of unification and supersymmetry in particle physics in the theory division at CERN, Geneva.

SHELDON L. GLASHOW is professor of physics at the Lyman Laboratory, Harvard University. He shared the Nobel Prize for Physics in 1979 for his theoretical work on the unification of the electromagnetic and weak nuclear forces.

GEOFFREY HALL is in the High-Energy Nuclear Physics Group at Imperial College, London.

JOHN LAWSON is currently working on the physics of particle beams at the SERC's Rutherford Appleton Laboratory, Oxfordshire.

HARRY J. LIPKIN is professor of theoretical physics at the Weizmann Institute of Science, Rehovot, Israel.

PAUL MATTHEWS is currently at the Department of Applied Mathematics and Theoretical Physics, Cambridge University.

DAVID PATTERSON is a science programme producer with BBC TV, Kensington House, London.

MARTIN L. PERL is professor at the Stanford Linear Accelerator Center in California. He played an important role there in the discovery of the particle known as the tau lepton.

PETER ROBERTSON, an editor of the *Australian Journal of Physics*, is currently with the Commonwealth Scientific and Industrial Research Organisation, Australia.

ROBERT WALGATE is a former science news editor of *New Scientist* and is now chief European correspondent of *Nature*.

DAVID J. WALLACE is with the University of Edinburgh where he is Tait Professor of Mathematical Physics.

ANDREW WATSON is currently at the Gulf Polytechnic, Isa Town, Bahrain.

VICTOR F. WEISSKOPF is professor of physics in the Department of Physics at the Massachusetts Institute of Technology, Boston. He was Director of the European Organisation for Nuclear Research (CERN) in Geneva, from 1961 to 1965. He is author of *Knowledge and Wonder* (MIT Press, 1979).

BILL WILLIS is in the experimental physics division at CERN, Geneva.

Foreword

In October 1957 the Soviet Union launched Sputnik 1, the world's first artificial satellite. A month later, a small and enthusiastic team of journalists and scientists launched *New Scientist*. Ever since then, with a few interruptions, *New Scientist* has provided a weekly dose of news from the world of science and technology.

Over the years, the magazine has reported research as it happened. Sometimes the findings that we described turned out to be less enduring than their discoverers first thought: sometimes they went on to win Nobel prizes for the scientists involved. Theories that first provoked severe scepticism became established wisdom. Established wisdom was cast aside.

These guides bring together that "history in the making". In these pages you will find more than a scientific account of a particular subject. You will find the personalities and the problems; the excitement of discovery and the disappointment of wrong turnings and the frustration of delays. I hope, too, that you will find something of the excitement that scientists feel as they push back the frontiers of knowledge.

Science writing does not stand still any more than does science. So the past quarter of a century has seen changes in the way in which *New Scientist* has presented its message. You will find those changes reflected here. Thus this guide is more than a collection of articles carefully plucked from the many millions of words that have appeared over the years. It is a record of science in action.

Michael Kenward
Editor
New Scientist

Introduction

The world we inhabit contains a wonderful diversity, not only in the widely varying forms of life that occupy every possible habitat, but in the materials that form those habitats. Between the shiny blackness of a lump of coal, which has taken millennia to form, and the ephemeral whiteness of foam on the sea, which disappears as quickly as it materialises, there lies a vast range of colours, textures and other properties, all of which describe some form of matter on this planet. It seems quite remarkable that we should dare to wonder if this great variety is based on a few simple building bricks; it is even more remarkable that we should find that this is indeed so.

The alchemists of the Middle Ages were possibly the first serious investigators of the nature of matter; we think of them as the fathers of chemistry. But it was the chemists of the 18th and 19th centuries – such as Antoine Lavoisier and Joseph Priestley – whose experiments probed into matter sufficiently deeply to lead John Dalton in 1808 to his atomic theory: that each chemical element is composed of identical "elementary particles", which differ from one element to the next. Thus the characteristic properties of a pure element – be it a gas such as oxygen, or a metal such as gold, or a powder like one form of sulphur – are determined by the nature of its atoms.

We still refer to the "chemical" elements today, but the study of the nature of matter has passed from the hands of chemists to the physicists. The idea that chemistry is "what things are made of" and that physics is "how things work" has become blurred. In studying how atoms "work", physicists in the 20th century have learned much about what atoms are "made of", and this has led to a remarkable new understanding of the fundamental nature of matter.

The alchemist of the 1980s is the "particle physicist" – the physicist who investigates the elementary particles of matter. But ask such a physicist what things are made of, and you will find that these elementary particles are not the atoms that Dalton described, but objects called "quarks" and "leptons". Atoms, the elements, this book, you, me, indeed our whole Universe, are constructed from these two building bricks, as far as we can tell.

Now, a child's building set with only two kinds of brick, say square and oblong, might seem a little limited. But if the set contains 100 bricks or so, we could use it to build a variety of structures; after all, very different edifices, from a garden wall to a railway viaduct, can all be built from the same basic oblong brick. So having only two types of basic unit to build the variety of the physical world is perhaps not so strange. And to be fair, I should admit that each of the basic units – the quark and the lepton – come in two different forms, so that we do in fact have four types of brick to build the world about us.

But what *are* quarks and leptons? The quarks are nothing more than the constituents of larger, more familiar particles – the protons and neutrons of the atomic nucleus. Forty years ago, physicists would in fact have told you that protons and neutrons were two of the "elementary particles" from which all matter is made. It was in the intervening years that experiments revealed a deeper structure: protons and neutrons appear to be built from still smaller particles which we call quarks.

What of the leptons? The best-known lepton is the electron, the carrier of electric current, the negatively charged particle that occupies the outer regions of an atom, orbiting the central nucleus rather as a planet orbits the Sun (but with some subtle and far-reaching differences). It is the electrons that take part in the chemical interactions between the "chemical elements"; in a sense the electrons determine the chemical nature as well as the physical properties of the elements. Copper conducts electricity because certain electrons move easily between its atoms; sodium is reactive because its outermost electron craves union with an atom of some other element, such as chlorine.

So much for the electron. The second type of lepton that we need to build the world about us is less familiar, and much more elusive. It is the neutrino: an electrically neutral, probably massless particle, which nevertheless plays an essential role in forming the world we inhabit. The neutrino, unlike the electron and the quarks, does not occupy a place in the atom. Instead it is created in processes in which one kind of quark turns into the other kind. At a more familiar level this means that neutrinos are involved in the transmutation of a proton into a neutron, or vice versa. The modern particle physicist is indeed the counterpart of the medieval alchemist who wished to transmute base metal to gold.

One of the most basic reactions in which neutrinos participate occurs in the first stage of the remarkable cycle that fuels the Sun. The solar furnace "burns" hydrogen nuclei (single protons) turning them into helium nuclei (each with two protons and two neutrons) with a net release of energy. Some of this energy is carried away by the radiation that eventually reaches Earth as life-giving light and heat. The rest is carried by invisible neutrinos, released when a positive proton changes into a neutral neutron. At the same time another lepton is released: not in this case the negative electron with which we are familiar, but a very similar particle with positive charge instead. This particle is known as the positron; it is not another kind of lepton, but rather a "mirror-image" of the electron. The positron has the same mass as the electron but positive charge. It is in fact the "anti-particle" of the electron.

The positron and the electron are alike in mass, but differ in electric charge. Once formed in a process such as the transmutation of a proton, the positron would live for ever, but for one fact. When electron and positron meet they soon "annihilate" each other. In other words, their properties cancel out, to leave only their total mass, which is converted into pure energy, just as Albert Einstein described in his famous equation, $E = mc^2$, where E is energy, m is mass and c is the velocity of light. The existence of the positron was predicted in the late 1920s by Paul Dirac, a theorist at Cambridge University. In writing down a theory of the electron that fulfilled the requirements of Einstein's theory of special relativity, Dirac came across the necessity for the positron.

I said earlier that we find we need two types of quark and two types of lepton to build the world about us. I neglected to mention then that these elementary particles must also exist in their anti-particle form. There must be anti-quarks forming anti-protons and anti-neutrons, in addition to anti-neutrinos and anti-electrons (positrons); this is all a consequence of Dirac's theory. Experiments have shown that these anti-particles do all exist; so why not anti-*matter*, with anti-electrons orbiting anti-nuclei built from anti-quarks? And if anti-matter can be formed why has it not annihilated with all the matter to leave only energy, rather than our evidently substantial material world?

This is the point at which "particle physics" becomes more than just a study of the building bricks of matter, and begins to touch upon our conceptions of the whole Universe. For although anti-matter can exist in theory, there is no evidence that it does exist anywhere on Earth, in the Solar System, in the Galaxy or anywhere nearby in the Universe. It is more difficult to be categorical about whether or not there is indeed any anti-matter at all in the Universe, and it is in dealing with problems like this that particle physics must join hands with astrophysics and cosmology. Particle physics becomes a study not simply of the elementary particles that Dalton believed he had defined, but of the necessary parts to build a Universe.

It is in this context that we should view the work of particle physics over the past 20 years or so, which is spanned by the chapters of this book. The experiments described create novel, short-lived forms of matter, which seem to require new kinds of quark and new kinds of lepton, and which at first sight seem to have little to do with reality. After all, we need only two types of quark and two kinds of lepton to build the world in which we live. But the physicists have found that some of these extra, "unnecessary" particles are created naturally. They are produced in the upper atmosphere in ultra high-energy collisions between cosmic rays from outer space and atmospheric nuclei.

The cosmic rays are energetic particles, mainly protons, but also heavier, more complex nuclei, such as iron. Where they come from is still unknown, although they seem to originate from beyond our Galaxy, and somewhere they are subjected to forces that accelerate them to extreme energies.

In their Earth-bound experiments, particle physicists cannot match the accelerating power of the mechanism that propels the cosmic rays. But they can perform their investigations under more carefully controlled conditions than when they study the random arrivals of the cosmic radiation. In this way, the experiments in laboratories that reveal the need for more quarks and more leptons echo the message of the cosmic rays, and cast light on the workings of the Universe beyond the confines of the Solar System. In particular, these studies reveal the nature of the forces that mould the Universe; the forces that create the forms we recognise from the basic building bricks of the quarks and leptons.

If the quarks and leptons are the bricks of our Universe, then the forces are the mortar that holds the bricks together and gives the construction its shape. At present we know of four forces that operate in different ways. There is gravity, which keeps our feet on the ground, controls the celestial clockwork of the planets, and binds thousands upon millions of stars within the Milky Way. Gravity operates across the immense distances of the Universe. Then there is the electromagnetic force which holds the negative electrons to the positive nuclei, and which underlies much of chemistry and therefore much of life. The remaining two forces act only within the confines of the nucleus. The strong nuclear force binds the quarks together within each individual proton and neutron. It is also responsible, though less directly, for holding protons and neutrons together. Lastly, there is the weak nuclear force, which controls such processes as the transmutation of protons to neutrons. Thus the weak force allows the Sun to burn.

We can see that each force is in its own way quite fundamental to our existence. Moreover, it is the precise nature of the forces that makes the Universe the place we observe. But particle physicists seek more than just an understanding of each force in its own terms. They hope one day to come to a single theory that encompasses all four forces, and thereby all the diversity of our Universe.

Such a goal may still be a long way off, but some success has already come with the so-called electroweak theory, which "unifies" electromagnetism with the weak nuclear processes. These unified theories show that the forces begin to resemble each other only as we go to higher and higher energies. As in our studies of the building bricks – the leptons and quarks – we find we have to go beyond processes here on Earth to learn more about the workings of our Universe. Thus the particle physicists have turned to study the Universe itself, treating it as a huge laboratory in which a number of experiments are in progress.

One "experiment" that astrophysicists can only dimly perceive in their studies of the radiant energy of the Universe, is the most important of all, the big bang with which cosmologists believe the Universe began. In the minute fractions of a second after the big bang, the Universe would have been so tremendously hot and energetic that it would have created conditions the particle physicists can only dream on. But today's experiments on Earth, at relatively feeble energies (and temperatures) compared with the

big bang, guide theory and provide it with the direction in which
to look back at the big bang.

Were the four forces then as one? Was the whole Universe a
boiling soup of quarks, moving around too fast to condense into
the protons and neutrons we observe now? How many kinds of
quark and lepton were there? Why indeed does the Universe seem
to require more of these building blocks than are required for life
on Earth?

The answers to these and other questions remain unanswered.
But as the chapters of this book reveal, particle physics has come
to mean much more than the study of the "elementary particles".
The particle physicist of today is in some senses the astrophysicist
of tomorrow. Both seek a better understanding of the complete
Universe. Who knows, the differences between them may disappear in the next century, as we come to learn still more about
building a universe.

PART ONE

Alpha to Omega

By the 1930s a new breed of scientist – the nuclear physicist – had appeared to have successfully unlocked the secrets of the atomic nucleus. The first decades of the 20th century had revealed not only the existence of atoms, which at the turn of the century some physicists had still doubted; they had also witnessed the elucidation of the atom's structure, in which tiny electrons orbit a central, heavy nucleus that contains most of the mass. The year 1932 proved pivotal in the growth of our understanding of the nature of the nucleus itself, as Peter Robertson shows in Chapter 1. It saw the discovery of the neutron – the proton's neutral companion in the nucleus – and of deuterium, or "heavy hydrogen", in which a neutron binds with the single proton of the ordinary hydrogen nucleus.

Two practical developments in 1932 were also to prove crucial to particle physics, which began to grow as a branch of nuclear physics. These were the first cyclic particle accelerator (the cyclotron) and the counter-controlled cloud chamber. This latter was a device in which cosmic rays could trigger cameras just as soon as the rays had passed through the chamber. In a sense the particles in the cosmic radiation photographed themselves, recording the trails of cloud-like droplets that they had left behind. The detector figured prominently in the discovery in cosmic rays of a number of new particles, particularly in the years after the Second World War. These new particles upset the confident picture established in the 1930s, in which protons, neutrons, electrons and neutrinos were sufficient to build all matter. And with the coming of powerful new accelerators in the 1950s, the discoveries that had been made in the collisions of cosmic rays in the atmosphere, were confirmed and amplified in man-made collisions between protons and other particles. Nature proved to have more to offer than the founding fathers of nuclear physics had ever suspected.

Part Seven will deal with the development of the accelerators and the detectors needed to capitalise on the new worlds opened up by the big machines; Part One will concentrate instead on the impact of the new discoveries on the study of "elementary particles".

By the early 1960s, as Victor Weisskopf describes in Chapter 2, a host of new "elementary particles" had been found. Many of these were assigned a new property, called strangeness, which served to distinguish particles and the way they behave, much as the property of electric charge determines the differing behaviour of protons and neutrons. (Weisskopf preferred to use the term "hypercharge" to label these particles when he wrote the article in 1962; now "strangeness" has become the accepted term.)

This seemingly arbitrary labelling of the new particles in fact provided the clue to establishing some degree of order among the growing chaos of the 1960s. Paul Matthews explains in Chapter 3 how theoretical physicists used mathematical ideas of symmetry to create a modern-day "periodic table". Just as in the 19th century Dmitri Mendeleev had brought order to the chemical elements, so theorists Murray Gell-Mann and Yuval Ne'eman discovered order in the "elementary particles" of the 1960s. Their scheme was indeed powerful enough to predict the existence of yet another new particle – the famed omega-minus, or Ω^-; its discovery in 1964 marked a watershed in particle physics, as Part Two will reveal.

1

The birth of nuclear physics

PETER ROBERTSON

In February 1932, James Chadwick discovered the neutron. This was the first in a series of remarkable developments that all fell within a single year.

It is often difficult to date the birth of a particular subject in science. Historians of science keep shifting ground, changing the relative importance they place on major discoveries. Nuclear physics is a case in point. Some might like to mark the subject's beginning with the discovery of radioactivity by Henri Becquerel just before the turn of the century. Others might consider the discovery of the atomic nucleus by Ernest Rutherford in 1911 to be decisive. In any case, all would agree that 1932 was a year of great importance – the year which saw the discoveries of the neutron, the positron and the deuteron, together with the completion of the first particle accelerators.

In 1932 the time was ripe for experimental physicists. During the 1920s theoretical physics had developed rapidly. Niels Bohr's theory of the atom, put forward in 1913, had come to its logical conclusion with the foundation of quantum mechanics in 1925–26. This new theoretical framework cleared the way for an understanding of many long-standing problems in atomic physics and the structure of matter. For more than a decade theory had prevailed over experiment, but by the late 1920s a feeling arose that a new era of experimental discovery and technical innovation was needed to take full advantage of the progress made in theoretical physics.

During this same period, research in physics underwent a broad demographic change. Whereas the atomic theory had been developed primarily in Copenhagen, Cambridge and Göttingen, by the late 1920s major research centres began to emerge in the

United States and elsewhere in Europe around such figures as Enrico Fermi in Rome and Iréne and Frédéric Joliot-Curie in Paris. Physicists at these new centres began casting around for uncharted areas to explore. The study of the nucleus and the phenomenon of cosmic rays seemed to be among the most promising.

The most important event in physics in 1932 came, however, from the well-established Cavendish Laboratory at Cambridge, under the direction of Rutherford. James Chadwick's discovery of the neutron was the culmination of a long and complex chain of events. At that time physicists had regarded the atomic nucleus as consisting both of heavy positively charged protons (hydrogen nuclei) and relatively light negative electrons – that is, in addition to the electrons orbiting the nucleus. The protons were thought to account for practically all the atomic weight, while the electrons were present in sufficient numbers to neutralise just enough protons to give the observed charge on the nucleus, which has the same magnitude as the total charge of orbiting electrons, and is called the atomic charge. This simple hypothesis seemed plausible as the discovery of beta decay at the turn of the century had shown that nuclei can emit electrons, and other work indicated that the atomic weights were fairly close to integral multiples of the proton mass. And yet for all the appeal of this straightforward picture difficulties arose, the most serious being the problem of nuclear stability. According to the so-called uncertainty principle that Werner Heisenberg had put forward in 1927, confining an electron in a volume as small as the nucleus would mean a large uncertainty in its momentum, which in turn would mean that the electron would be far too energetic to remain in the nucleus for more than a fraction of a second.

The solution to the problems did not come from the theorists. The key to nuclear structure had in fact already been proposed much earlier by Rutherford. In his Bakerian Lecture of 1920, Rutherford predicted the existence of a neutral particle, with roughly the same mass as the proton, but consisting of a kind of hydrogen atom in which the single electron had fallen into the solitary proton in the nucleus. "The existence of such atoms seems almost necessary to explain the building up of the nuclei of the heavy elements", claimed Rutherford (*Proceedings of the Royal Society*, vol. A97, p. 374). During the early 1920s researchers at the Cavendish Laboratory attempted to detect the formation of the hypothetical "neutron" by passing a strong current through a hydrogen discharge tube, but without success.

Then, in 1930, Walther Bothe and H. Becker working in Berlin reported a curious result. If they bombarded beryllium with alpha particles (helium nuclei) they found that the beryllium emitted a neutral radiation of great penetrating power, which the two German physicists understood to be gamma rays – high-energy electromagnetic radiation. Further, it seemed that these gamma rays had an energy nearly 10 times higher than the incident alphas, clearly a doubtful conclusion. In 1932, the Joliot-Curies reported an even more surprising result. The penetrating radiation emitted from beryllium could eject an intense stream of protons from a paraffin absorber – a phenomenon the researchers interpreted as a kind of scattering of the protons by the gamma radiation, but with apparently more than a million times the likelihood of any similar scattering process then known.

When the Joliot-Curies' paper arrived at the Cavendish Laboratory late in January 1932, Rutherford reportedly burst out with "I don't believe it". Chadwick did not believe it either. After a few days of inspired work, he had repeated the same experiment and felt convinced that the strange "gamma radiation" was none other than Rutherford's hypothetical neutron. On 17 February 1932, Chadwick despatched a letter to *Nature* (vol. 129, p. 312) that began with a discussion of the work by the Joliot-Curies and concluded: "The difficulties disappear, however, if it be assumed that the radiation consists of particles of mass 1 and charge 0, or neutrons."

The discovery of the neutron was accepted immediately. With the neutron in the picture it was no longer necessary to assume the presence of electrons in the atomic nucleus, and so to doubt the validity of quantum mechanics in describing the nucleus. The neutron not only opened up the study of nuclear structure and the nature of the nuclear forces, but also, with its large mass and zero charge, provided a powerful tool for penetrating and transforming the nucleus.

The neutron added special significance to another discovery made in the same year. The day before Chadwick sent his letter to *Nature*, the American journal *Physical Review* received a paper from Harold Urey, Ferdinand Brickwedde and G. M. Murphy reporting the identification of a hydrogen isotope of mass 2 – that is, a nucleus with the same charge but twice the mass of normal hydrogen. This discovery ended years of speculation about the existence of this isotope.

As early as 1919 researchers in Frankfurt had investigated

whether hydrogen might be a mixture of two isotopes, but their technique did not concentrate the heavier isotope sufficiently for its detection. The vital clue came in the late 1920s. Improved measurements revealed a small discrepancy between the atomic weight of hydrogen determined by chemical means and the value Francis Aston found with his mass spectrograph, an instrument that measures the deflection of ions moving through a combination of magnetic and electric fields. Raymond Birge and D. H. Menzel, two physicists at Berkeley, California, pointed out that this discrepancy could be explained if samples of hydrogen contained an isotope of mass 2 in the proportion of about one part in 4500. Urey at Columbia University decided to look for "heavy hydrogen" using a straight-forward fractional distillation of hydrogen near its triple point – the combination of temperature and pressure at which hydrogen exists as liquid and gas. Urey's belief was that the heavier isotope, being less likely to evaporate, would concentrate in the residue making it easier to detect. Ferdinand Brickwedde at the US Bureau of Standards prepared samples of different concentration which Urey and his research assistant Murphy analysed spectrally. They found that some of the strongest spectral lines expected for deuterium – members of the so-called Balmer series – were plainly evident, exactly in the positions calculated, proving without doubt the presence of the new isotope.

The discovery of deuterium, or more particularly its nucleus the deuteron, had important implications for nuclear physics. The deuteron with mass 2, but charge 1 as on the proton, consists of a proton bound with a neutron. Chadwick, in his second publication on the neutron in June 1932, pointed out that the deuteron, like the alpha particle (two protons and two neutrons), probably formed an essential structural unit of nuclei. But of greater significance than this notion of pairing in nuclear structure was the information the deuteron promised to yield on the nuclear force. Apart from normal hydrogen, all other nuclei consist of three or more tightly packed nucleons (protons and neutrons) and an exact analytical treatment of the dynamics of these many-bodied systems is not possible. So, the simple combination of neutron and proton in the deuteron offers an ideal test case for studying the general properties of the strong interaction which binds all nuclei. The deuteron thus played a central role in nuclear physics in the 1930s, in much the same way that the hydrogen atom had been the empirical proving ground for atomic physics in the previous decade.

Physicists barely had time to study further the neutron and deuteron before the announcement in September 1932 of the discovery of another new particle.

Carl D. Anderson at the California Institute of Technology had been studying the tracks that cosmic rays produced in a cloud chamber – an enclosure of gas in which electrically-charged particles produce trails of droplets of liquid as the gas is rapidly expanded. Anderson found compelling evidence for particles with a mass similar to an electron's, but with an equal and opposite

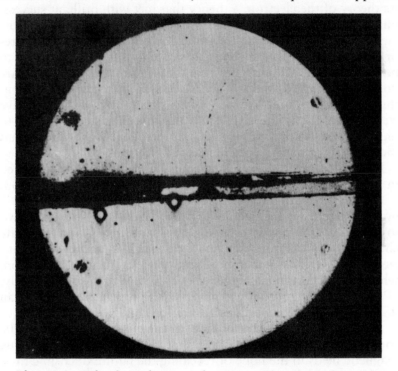

Plate 1.1 *The first glimpse of anti-matter. In August 1932, Carl Anderson observed in his cloud chamber the track of a particle with the same mass as the electron, but positive charge – in other words an anti-electron, or positron. In this picture the positron is travelling upwards, losing energy on passing through the lead plate, so that the curvature of its path increases. (Credit: C. Anderson Caltech,* Physical Review)

charge – in other words a positive electron or "positron". Initially, physicists greeted this discovery more with puzzlement than enthusiasm. The existence of both the neutron and deuteron had at least been suspected for a decade or more, and each had removed a number of obstacles to the progress of nuclear physics. But the positron was not really required; indeed it complicated the picture that all matter comprises the three elementary particles – protons, neutrons and electrons.

At the time of Anderson's discovery few physicists realised that a theorist had already predicted the existence of the positron. In 1928 Paul Dirac, working at Cambridge, published his relativistic theory of the electron, a theory that met with brilliant success, but which was plagued by an intractable problem. Dirac's formalism predicted that the electron should not only take up states of positive energy, but also negative energy states that seemed to be non-physical. Dirac argued that these negative energy states are normally occupied, though occasionally a state can empty out, thus creating a "hole", which will be associated with a positive electric charge, as the electron is negative. Physicists sought to identify these holes with protons, but Hermann Weyl showed that the holes must have the same mass as electrons.

Dirac then made the obvious connection. In a paper published in September 1931 (*Proceedings of the Royal Society*, vol. A133, p. 60), Dirac referred to the question of negative energy states. "A hole, if there were one, would be a new kind of particle, unknown to experimental physics, having the same mass and opposite charge to an electron. We may call such a particle an antielectron." At the same time Dirac made a further prediction, which would take much longer (24 years) to verify experimentally. "Presumably the protons will have their own negative states . . . an unoccupied one appearing as an anti-proton."

Anderson had not been looking for the positron. Instead, in the spring of 1932, he began subjecting cloud chambers to strong magnetic fields in the hope of revealing the nature of cosmic rays. To this end, Anderson set about measuring the energy spectrum of the secondary electrons produced by the incoming cosmic rays, and he routinely collected and examined thousands of photographs of the tracks curved by the magnetic field in the cloud chamber. Many of these tracks revealed positively charged particles, but several in particular seemed far too light to be either protons or alpha particles. These tracks could, in principle, have been made by negative electrons moving in the opposite direction (from above) through the cloud chamber – rather than by positive particles coming from below. To settle the question Anderson inserted a thin lead plate horizontally in the cloud chamber and, on 2 August 1932, he obtained a clear and unambiguous photograph of a light positively charged particle losing energy as it passed upwards through the plate. Anderson sent a short letter to *Science*, followed by a detailed paper in the *Physical Review* (vol. 43, p. 492) describing his discovery. "It is concluded therefore, that the magnitude of the charge of the positive electron which we shall henceforth contract to positron is very probably equal to that of a free negative electron which from symmetry considerations would then be called a negatron."

Anderson's general conclusions on the properties of the positron were correct, but his explanation of the mechanism for its production, similar to his proposed nomenclature for the electron, did not fare so well. Anderson believed the primary cosmic ray struck a neutron within a nucleus causing its disintegration into a positron and a negative proton, and he urged physicists to search for evidence of this negative or anti-proton, unaware that Dirac had already predicted its existence over a year earlier. The connection between the positron and Dirac's theory, as well as the correct explanation of its production mechanism, were provided shortly after by P. M. S. Blackett and G. Occhialini. In the summer of 1932 these two physicists at the Cavendish Laboratory had developed a new and more efficient automatic cloud chamber, and this enabled them to verify the existence of positrons within days of Anderson's announcement. Blackett and Occhialini suggested that electrons and positrons were created in pairs directly from high-energy gamma rays passing close to atomic nuclei. This hypothesis was not only consistent with Dirac's theory, but also provided a beautiful confirmation of its predictions.

It later became clear that several other groups studying cosmic rays had photographed the positron earlier but had failed to notice the distinctive tracks. The luckless Joliot-Curies, having already narrowly missed the discovery of the neutron, had also observed positron tracks late in 1931 during their study of the disintegration of beryllium nuclei by alpha particles. But they had dismissed these tracks as representing stray electrons moving toward the beryllium source. In a similar vein, Blackett once remarked that if Occhialini had not taken such a long holiday during the summer of 1932, they would have found the positron before Anderson did. While this may well have been true, the development of the automatic cloud chamber by Blackett and Occhialini probably did more to further the study of cosmic rays than the discovery of the positron. Since the invention of the first cloud chamber by C. T. R. Wilson in 1895, there had been no way of knowing whether a cosmic ray had actually passed through the chamber, so that gathering useful photographs was a rather hit-or-miss affair. Blackett and Occhialini devised a method in which they placed detectors above and below the chamber so that the chamber's expansion would be triggered only by the arrival of a cosmic ray. In this way cosmic rays took their own photographs, a method some 50 times more efficient than the cloud chambers that Anderson and others used.

The discoveries of the neutron and positron doubled in one year the number of known elementary particles from two to four, and the detection of the deuteron initiated studies of the simplest composite nucleus. But there were in the same year equally important steps forward in the techniques and instrumentation of nuclear physics. Progress in nuclear physics to this point had been heavily constrained by the almost complete reliance on naturally occurring radioactive sources. The most prized source, radium, at the going rate of about $100 000 a gram, was too expensive for most laboratories to use for providing a sufficiently intense beam of radiation. Besides, the alpha particles that radium emitted possessed enough energy to penetrate and disintegrate only the lightest elements.

Physicists soon realised that they needed to build some kind of device to accelerate copious amounts of charged particles to higher energies. The obvious way was through creating large electric fields – "a million volts in a soap-box" was Rutherford's phrase. Researchers tried all sorts of possibilities during the 1920s, but most techniques were limited by corona discharge or the break-down of insulation before they reached sufficiently high voltages. One foolhardy attempt to harness the high voltages in storm clouds resulted in a physicist losing his life, stuck by lightning.

Events took a turn for the better in 1928 when the Russian physicist George Gamow and, independently, Edward Condon and Ronald Gurney in the US showed that charged particles of fairly low energy did have a reasonable probability of penetrating a nucleus and causing its disintegration. Encouraged by these theoretical considerations, John Cockroft and E. T. S. Walton at the Cavendish Laboratory decided to build a proton accelerator employing the well-tried technique of the voltage multiplier. In

Plate 1.2 *John Cockcroft and E. T. S. Walton accelerated protons to 700 kilovolts using this four-stage voltage multiplier, and produced the first controlled nuclear transmutations early in 1932. But their technique was superseded by the circular accelerator – the cyclotron – developed by Ernest Lawrence, which first achieved 1 million volts in February 1932. (Credit: Cavendish Laboratory)*

this device a four-stage arrangement of electrical capacitors and vacuum tubes multiplied a relatively small input voltage up to a potential of almost 700 kilovolts at a discharge tube in which protons were accelerated. Early in 1932 Cockroft and Walton placed a lithium target at the end of the discharge tube, turned the proton beam on, and saw on a zinc sulphide screen the familiar scintillations characteristic of alpha particles. Protons had cleaved lithium nuclei into two alpha particles. This was the first completely controlled nuclear transformation and marked the first significant step in the history of accelerator physics.

Despite this early success, the Cockroft–Walton machine was rapidly superseded by the invention of a circular magnetic accelerator – the cyclotron – which avoided most of the problems inherent in the generation of high voltages. Early in 1929 an American physicist, Ernest Lawrence, prompted by an idea in a paper by Norwegian electrical engineer Rolf Widerøe, realised that charged particles could be accelerated in easy stages if they were kept moving in circular paths by a magnetic field. An alternating voltage could give the particles a kick every half revolution, the cumulative effect being that the particles would gain energy, moving in ever widening circles. The beauty lay in the idea of multiple acceleration, which meant that only relatively small operating voltages were required. Lawrence constructed a small working model and his graduate student M. Stanley Livingston showed that the gadget actually worked by producing a beam of 80 kV protons. Rather than attempting any experimental nuclear studies at this stage, Lawrence and Livingston opted to forge ahead and build a larger cyclotron with the magnet pole enlarged from 4 to 11 inches. However, the beam of this larger machine for some undetermined reason defocused and smashed into the walls of the acceleration chamber well before the protons reached the maximum energy possible and Lawrence and Livingston spent months tinkering before a series of trial-and-error changes finally restored the beam focus. In February 1932 they coaxed the machine past the magic mark of 1 MV.

News of the success of Cockroft and Walton in disintegrating lithium nuclei with voltages as low as 400 kV reached Lawrence's laboratory at Berkeley in California shortly afterwards. The result was quickly confirmed in the cyclotron and soon a variety of elements were being bombarded by the energetic proton beam. In March 1933 Lawrence received a small sample of heavy water from the chemist Gilbert Lewis, only a little over a year after

Urey's discovery of deuterium, and he used this to produce a deuteron beam with the cyclotron. The energetic deuterons initiated a wide range of new nuclear reactions, many of which yielded neutrons among the products.

The cyclotron at Berkeley heralded a new era in nuclear physics. By 1936 there were about 20 cyclotrons either operating or under construction throughout the world, each complicated machine requiring, by the standards of the day, large teams of physicists, engineers and technicians and large capital expenditures. Big science had begun. A generation of physicists turned their collective talents to the glamorous new field of nuclear physics, seeking explanations of the liveliest issue in physics, the structure and properties of the nucleus.

The breakthroughs of 1932 were of crucial importance to the early development of nuclear physics, and perhaps their overall significance can best be gauged by the rich harvest of Nobel prizes they yielded for the principal figures. Chadwick received the prize in physics for his discovery of the neutron and Anderson followed suit in 1936 for his discovery of the positron. Urey's discovery of deuterium earned him the prize in chemistry as early as 1934. The machine builders also fared well, Lawrence winning the physics prize in 1939 for his invention of the cyclotron, although it was not until 1951 that Cockroft and Walton received recognition in Stockholm for their work. Blackett's prize in 1948 was awarded largely for his development of the automatic cloud chamber in 1932 – a miraculous year for physics.

18 February, 1982

2

The world within the proton

VICTOR WEISSKOPF

Experiments bring to light an unexpected richness in the physics of the proton and the nuclear force.

Since the discovery at the turn of the century that atoms are not indivisible or immutable, the central problem in physics has been to understand the nature of the atom and its constituents in greater and greater detail. To the outsider, workers in this field sometimes appear to be an exclusive brotherhood who deal in occult symbols and make inexplicable experiments with complicated machinery. In fact what we are doing is quite straightforward; it is difficult only because Nature will not yield her innermost secrets readily. A man who wants to know what makes a clock tick has to take it to pieces; to find out how matter is composed and how it works we have to take it to pieces. It is a fundamental consequence of quantum laws that, as the pieces get smaller, we have to smash them with ever more powerful electromagnetic sledgehammers – our particle accelerators.

Historically, we can mark three stages in the exploration of sub-atomic matter during this century. First, there was the elucidation of the structure of the atom, as a cloud of electrons arranged in shells, according to the laws of quantum mechanics, around a central heavy nucleus. Next, the structure of the nucleus unfolded – and it is seldom appreciated what a great source of intellectual satisfaction there was in the discovery that the nucleus too contained shells of particles obeying quantum laws. Here the particles are nucleons – protons and neutrons. The third stage is the investigation of the structure of the nucleons themselves.

There is nothing accidental about this course of events. The three stages deal with three distinct sub-atomic worlds within worlds. To disturb the electron structure of an atom requires energy of only a few electron volts; the forces concerned are electromagnetic, and in the description of the atom the nucleus is simply a small, heavy, electrically charged unit. This "electron-volt" world is the world of our daily life: the physical and chemical properties of materials, our chemical fuels and the complex chemistry of life all depend on "electron-volt" phenomena, and with the advent of the quantum mechanical concept of the atom much has become explicable.

Much higher energy is needed to break up a nucleus – energy of the order of a million electron volts (MeV). In this "MeV" world, the physicist encounters a new force stronger than the electromagnetic forces of the familiar world – the nuclear force which binds the nucleons together. The "MeV" world is the world of nuclear reactors and nuclear bombs. More fundamentally, it is the world of the Sun and the stars, and its phenomena include their nuclear fire and the synthesis of the familiar chemical elements from the primeval hydrogen of the Universe. In fact, given hydrogen, and our knowledge of "electron-volt" and "MeV" physics, we can in principle explain the creation of our world and all its colour and life. That is the sum of the accomplishment of science in the first half of the 20th century.

"Given hydrogen" that is where we take up the story and inquire into the nature of the elementary particles, the electron and especially the nucleon, of which hydrogen is composed. To break up a nucleon requires energy of the order of a thousand million (10^9) electron volts (GeV). These marked jumps in energy between the various stages of our quest are fortunate for the experimentalist because they prevent phenomena at one level interfering with observations at another. It is only since the Second World War that we have been able to create the "GeV" world in our laboratories (Part Seven). Before that we had glimpses of its phenomena in the impacts of the extremely energetic particles of the cosmic rays. But the cosmic rays are diffuse, and it may be doubted whether the conditions prevailing at the moment when a strong beam of particles in a proton synchrotron hits a material target exist anywhere in the natural Universe. Yet in the puzzling new phenomena which we discover by shooting our powerful proton beams at material targets we hope to find further explanations of the natural world: the nature of the nuclear force, and its relationship with electromagnetism and gravity, and the mechanism of radioactivity.

Our "GeV" world is apparently more complicated than the tidy shell structures of the atom and the nucleus, but that may be only because we cannot yet draw a logically conclusive picture. At present all we can do is describe the phenomena and try to group them in suitable ways. Personally I favour three main groups: nuclear force phenomena, the anti-world, and the so-called "weak interactions".

Nuclear force phenomena. – If we jerk an electron, in a radio transmitter or a lamp filament, it gives off photons – quantum particles of electromagnetic radiation which can be regarded as manifestations of the electromagnetic field of the electron. Similarly, if we jerk a nucleon by hitting it with a high energy particle, it gives off particles: mainly pi-mesons and K-mesons. But the nuclear force is more complicated than the inverse square laws of electromagnetism and these particles have mass and sometimes electric charge. The K-mesons also possess another property which has been called "strangeness", but I prefer the term "hyper-charge". When a nucleon emits a charged pi-meson it changes from proton to neutron or vice versa, but when it loses a K-meson its "hypercharge" changes and it turns into something else – a lambda (Λ) or a sigma-particle (Σ). If a nucleon is struck so violently that it loses two K-mesons, it changes into a xi-particle (Ξ), which thus differs from the original nucleon by two units of hypercharge.

There are in fact eight particles which form a group differing from one another only by the addition or subtraction of charge or hypercharge: the proton and neutron, the neutral lambda, three sigma-particles (+, 0 and – charge), and two xi-particles (0 and +). To be sure, they differ in mass as well, but as mass is just a particular form of energy we do not attach too much importance to it. These particles are together known as baryons, and are listed along with other particles in Table 2.1.

To complicate the picture, both the baryons and the mesons can appear in forms other than the ones we have mentioned so far. For example, there exist excited states of some of the eight baryons. One is called Y* (an excited state of the sigma particle); others are referred to as "higher resonances". Just as an ordinary atom, when disturbed, returns to its "ground state" by the emission of photons, such excited states quickly return to one or other of the eight fundamental baryon states, by the emission of pi-mesons. The mesons also seem to appear in excited states; we call them the rho- and omega-mesons in the case of the pi-meson, and the K*-meson in the case of the K-meson.

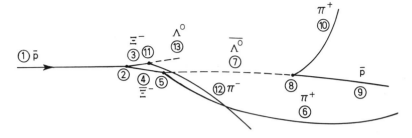

Figure 2.1 *The complex reaction in which the anti-xi-minus was discovered in the 81-cm liquid hydrogen bubble chamber at CERN. An anti-proton (1) collides with a proton (2). Both are annihilated but result in the formation of a negative xi (3) and the new anti-particle (4). The latter then decays (5), 10^{-10} s later, into a positive pion (6) and a neutral anti-lambda (7), which itself decays (8) into an anti-proton (9) and a positive pion (10). The negative xi decays (11) into a negative pion (12) and a neutral lambda (13).*

The anti-world. – The positron, the positively charged anti-electron, was discovered in the cosmic rays, but until the advent of the big machines it was not possible to confirm the theoretical prediction that for every heavier particle, too, there is an anti-particle. Now, the heavier anti-particles have been checked off one by one, from the anti-proton to the anti-xi-minus (see Figure 2.1).

When a particle meets its anti-particle the remarkable phenomenon of annihilation occurs: the matter is converted into energy. In electron-positron encounters the energy appears as gamma-rays, but when the heavier particles are annihilated the usual "currency" for the energy is pi- and K-mesons.

Weak interactions. – Nearly all the units of matter which I have mentioned are very short-lived by human time-scales – typically they decay after 10^{-10} s, which is tiresome for the experimentalist. Yet, in the "MeV" and "GeV" worlds of the nucleus their lives are very long: the nuclear "year", the time for a nucleon to orbit in the nucleus, is about 10^{-22} s. This widespread phenomenon of decay – also called "weak interaction" because the forces involved are very much less strong than the nuclear force – is the most curious aspect of these units of matter.

The neutron decays into a proton and an electron, and this process manifests itself when a radioactive nucleus throws out an electron (a beta-particle). But early studies of beta-radioactivity revealed that some energy and momentum was being lost in the process, which led to the prediction of an almost undetectable particle, the neutrino, which carried energy and momentum away with it. The existence of the neutrino has since been amply confirmed. But that was not all. Having invoked the neutrino to make sure that energy and momentum are properly conserved, we have now learned, in the famous "non-conservation of parity" experiments, that in beta-decay the electrons and neutrinos emitted have a bias in favour of one direction of spin rather than the other: the electrons spin anticlockwise about their motion, like a left-handed screw-thread.

Another puzzling discovery is that a lambda-particle can decay into a neutron and a pi-meson. As I have explained, a lambda-particle differs from a neutron by its property of hypercharge, and

Table 2.1 The particles known in 1962 and their characteristics

Type of particle	Symbol	Rest mass m_e	Measured mean lifetime in free space	Anti-particle
Photon				
Leptons	γ	0	—	—
Neutrino	v			
Electron	e^-	0	—	$\bar{v}$
Muon	μ^-	1	—	e^+
		206.9	2.2×10^{-6} s $\pm$ 1%	μ^+
Mesons				
π-mesons	π^+	273.2	2.6×10^{-8} s $\pm$ 2%	π^-
	π^0	264	$\sim 10^{-16}$ s	π^0
	π^-	273.2	2.6×10^{-8} s $\pm$ 2%	$\pi+$
K-mesons	K^+	966	1.22×10^{-8} s $\pm$ 1%	K^-
	K^0	975±	9×10^{-11} s $\pm$ 8%	K^0
Baryons				
Nucleons	p	1836.1	—	$\bar{p}$
	n	1838.6	17.3 min $\pm$ 13%	$\bar{n}$
Λ^0 hyperon	Λ^0	2182	2.7×10^{-10} s $\pm$ 6%	$\bar{\Lambda}^0$
Σ hyperons	Σ^+	2328	9×10^{-11} s $\pm$ 7%	$\bar{\Sigma}^-$
	Σ^-	2342±1	1.7×10^{-10} s $\pm$ 10%	$\bar{\Sigma}^+$
	Σ^0	2330±2	$<10^{-13}$ s	$\bar{\Sigma}^0$
Ξ hyperons	Ξ^-	2584±4	$\sim 10^{-9}$ s	$\bar{\Xi}^+$
	Ξ^0	~2584	$\sim 10^{-9}$ s	$\bar{\Xi}^0$

the pi-meson does not carry any hypercharge. So, in weak interactions, hypercharge can disappear, although ordinary charge is conserved.

The pi- and K-mesons also decay in many ways. For example, a pi-meson decays into an electron and a neutrino. I do not mean here the familiar electron of the television picture tube. To the astonishment of the physicist, it seems to be an electron two hundred times heavier than usual – the muon – which is a guise that the electron can, with baffling duplicity, adopt whenever there is mass or energy to spare.

It is possible to sum up the elementary particles by dividing them into three groups: (1) *baryons*, the heavy nucleons in their various forms; (2) *mesons*, pi and K, representing the nuclear force; and (3) *leptons*, the electrons, muons and neutrinos which figure in the weak interactions. But opinions differ and my description is arbitrary. For example, some physicists deny a real distinction between the nucleons and the mesons.

As I warned at the outset, we can describe events in the "GeV" world but not explain them. With the coming into operation of new and powerful machines in the early 1960s the phenomena

Plate 2.1   *CERN's Super Proton Synchrotron, showpiece of a GeV laboratory that takes us deep into the world within a proton. (Credit: CERN)*

have proved to be richer and more varied than expected. The near-monopoly of "GeV" equipment enjoyed by the United States has been successfully challenged by Europe and the Soviet Union. Our exploration is very costly: it is not just the initial cost of the accelerators themselves; peripheral equipment and instrumentation costs when the machine is running requires an annual outlay at least as much as the per annum cost of building the machine. I hope that the authorities in Europe will realise, if they want the large "GeV" laboratories to fulfil their promise, that money buys time and time wins discoveries about the fundamentals of matter which would otherwise go by default to later starters.

It might be comforting if I could say that an end is in sight. But even if all the sub-nucleon matters I have discussed are in due course resolved as neatly as the atomic and nuclear structures, who knows where we shall end, if we then begin to inquire into the structure of whatever we prove to be the units which make up the nucleon?

28 June 1962

3

Order out of sub-nuclear chaos

PAUL MATHEWS

Attempts to understand the variety of the so-called "elementary particles" reveal a new symmetry in nature and predict a new particle.

February 1964 should go down in the history of physics as the time when a new fundamental law was established.

The frontier of physics has lain, since the Second World War, in the field of sub-nuclear particles. There have been spectacular advances, each 5-year period producing a new and quite unexpected development. But, so far, each new stride has seemed like a step backwards in that it has brought ever-increasing complexity and confusion into a domain of physics where one had reason to expect the last word in precision and simplicity. This month, however, the tide is turning and a coherent pattern is emerging, to bring order and beauty into this sub-nuclear world.

By 1932 it had been established that atomic nuclei are made of protons and neutrons (nucleons), packed together into tiny spheres with radii of about 10^{-12} cm. It had by then been realised that very powerful, specifically nuclear, forces must operate between these particles in order to overcome the enormous electrical repulsion between the tightly packed, positively charged protons. In 1935 Hideki Yukawa suggested a possible mechanism for this force: that protons and neutrons interact by the exchange of yet another particle – the pi-meson or pion – in much the same way that rugger players interact by exchanging the ball. From the range of the force, he was able to predict the mass of the new particle. Very general considerations then implied that these pions should be produced in sufficiently energetic collisions between nucleons. The explosion of nuclear weapons in 1945 amply displayed the strength of the nuclear forces, but the confirmation of Yukawa's ideas was seriously delayed by the war. They were beautifully established in 1947, when C. F. Powell and his group at Bristol discovered pions in nuclear emulsions exposed to high-energy protons in cosmic rays.

At about the same time, C. D. Rochester and C. C. Butler (then of Manchester) found the first evidence for many more quite unexpected sub-nuclear particles. There followed a period of several glorious years of confusion, in which new particles seemed to appear almost monthly. By the late 1950s a well-defined picture had emerged.

Table 3.1 shows particles which interact through the strong nuclear forces. All these particles can be produced copiously in sufficiently energetic nuclear collisions (for example, in the proton beams of the big accelerators at CERN in Geneva and at Brookhaven, Long Island), As Table 2.1 showed, apart from the proton, the particles are unstable, with mean lives of about 10^{-10} s. In everyday terms this is a very short time, but on the nuclear time-scale it is very long. These apparently ephemeral particles, to a very good approximation, may be regarded as stable.

Table 3.1   The quantum numbers of the particles known in 1963

Name	Charge Q	Isospin I	I_3	Hypercharge Y	
Xi (Ξ)	0 −1	½	$+½$ $-½$	−1	
Sigma (Σ)	+1 0 −1	1	+1 0 −1	0	Spin-½ baryons
Lambda (Λ)	0	0	0	0	
Proton (p) Neutron (n)	+1 0	½	$+½$ $-½$	1	
Kaon (K)	+1 0	½	$+½$ $-½$	1	
Anti-kaon ($\bar{K}$)	0 −1	½	$+½$ $-½$	−1	Spin-0 mesons
Pion (π)	+1 0 −1	1	+1 0 −1	0	

Thus, there are two quite distinct types of nuclear force: the "strong interaction" which binds the nucleus together, and through which the particles are produced, and the disruptive "weak interaction" which causes these particles to disintegrate, or decay.

In his attempts to understand why the various particles exist and how they behave, the physicist searches for rules that govern them and particularly for physical quantities which remain unchanged in any interaction. For example, he knows that if two particles collide and undergo changes, the total energy of the system (which includes mass as a form of energy) is not altered, whatever the products of the interaction may be. Similarly, the net electric charge must be conserved. Another general rule can be deduced from the stability of the proton. If there were any chain of reactions whereby protons could transform into lighter particles, they would have done so by now, and the familiar materials of the world would not exist. The fact that this is not so shows that the total number of protons (or more massive particles) is unchanged in any process. This is known as the conservation of baryons (heavy particles).

It will be seen that the particles of Table 3.1 all have a "spin" ascribed to them (though in the case of the mesons shown the spin is zero). Spin is a kind of angular momentum, and just as angular momentum is conserved in the more familiar world of tops and planets, so in interactions between particles the total angular momentum, including the spin, remains unchanged.

To make further progress, it was necessary to suppose that the particles possessed other intrinsic attributes, beyond charge and spin, which also had to be conserved. The immediate analogies with objects in the macroscopic world were, however, exhausted and the more abstract notions of "isotopic spin" and "hypercharge" were introduced. Although we cannot visualise these properties, except in mathematical terms, they prove to be real enough in their effects.

We can see how they arise by considering two related features of the discovered particles.

Firstly, they appear in groups, or "multiplets", of particles of different charge but very nearly equal mass. Secondly, a number of the decay processes (for example, the xi-particle breaking up into a lambda-particle and a pion) involve only strongly interacting particles, yet they happen so slowly that they cannot take place through the strong interaction; there must, therefore, be some conservation law which prevents these decays occurring via the strong interactions, so that they have to go, much more slowly, via the weak interaction.

These features were explained by introducing the concept of isotopic spin, I. (This is now usually referred to simply as "isospin".)

It is mathematically analogous to ordinary spin. Since the orientation of the spin axis can vary, giving different components of isotopic spin (I_3) in some arbitrarily chosen direction, the isotopic spin affords us a set of related particles distinguished by the different values of I_3 (Figure 3.1). There is, in fact, a simple quantum rule that tells us that a particle of isotopic spin I (which can be an integral or half-integral number) occurs as a multiplet of particles having $2I+1$ members. Given this rule, and knowing the

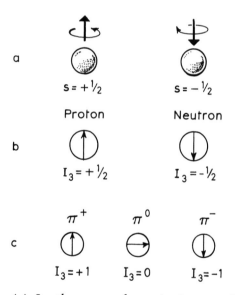

Figure 3.1 *(a) In the case of a spinning particle we can define its sense of spin about a particular direction — as given by a right-handed screw thread, for instance — and assign values for spin, s, accordingly. (b) We can carry this notion over to an abstract form of spin — isospin, I — which serves to distinguish similar particles, such as the proton and neutron. The component of isospin, I_3, is then different for the different particles. (c) In the case of the pi-meson (π), which has three charge states, I_3 takes on three values.*

number of observed charge states for each particle, we can assign values of I, as in Table 3.1.

It then turns out that a conservation law requiring that isotopic spin is conserved in strong interactions is precisely what is needed to forbid the unwanted fast decay reactions. The xi-particle, for example, having half-integral isotopic spin, cannot possibly decay into a lambda (zero isotopic spin) and a pion (integral isotopic spin) while conserving isotopic spin. It must, therefore, do so via the slower weak interactions.

We arrive at the concept of hypercharge, Y, when we notice that, if we subtract the isotopic spin component I_3 from the charge Q for each of the particles in Table 3.1, the difference is the same for all the particles in a multiplet. We can then take this number as an intrinsic property of the multiplet. In practice we take twice the difference: hypercharge, $Y=2$ $(Q-I_3)$. Since Q and I_3 are conserved in strong interactions, Y is also conserved. Indeed, the physical implications of the scheme are most easily seen by considering hypercharge conservation. Hypercharge is not conserved, for example, when the xi decays into a lambda and a pi.

All this was established in 1957. Our experimental story starts again in 1960, with the discovery that Table 3.1 contains only the unexcited forms of sub-nuclear matter, and that there is a whole spectrum of many tens of sub-nuclear particles (the masses of the particles being analogous to the energy levels of atomic or nuclear systems). The old notion that Table 3.1 is a list of "elementary particles" has completely broken down, and the question has switched to the one of discovering the conservation law which controls the pattern of the larger multiplets of particles thrown up by this new sub-nuclear spectroscopy. This is the problem which seems to have been solved.

It is fairly clear that the solution must lie in a generalisation of the idea of isotopic spin. The natural generalisation was from spin in three dimensions to spin in four dimensions, but this led nowhere.

The theoretical breakthrough came from the Japanese physicist Y. Ohnuki in a paper presented at the Rochester Conference on High Energy Physics in 1960. Ohnuki suggested that the correct thing to consider was not rotations, but the mathematically closely related notion of "unitary transformations". Charge conservation can be related to unitary transformations in one dimension; isotopic spin to two-dimensional unitary transformations. Ohnuki considered the generalisation to three dimensions. Hardly anyone took the slightest notice. But the notion was picked up by Abdus Salam at Imperial College, London, in conjunction with his ideas on gauge theories. He worked on it in London with J. C. Ward, and infected a research student Yuval Ne'eman with his ideas. Professor Murray Gell-Mann, who spent some time in London on leave of absence from Caltech, also became interested.

The requirements that Ohnuki's new "spins" should be conserved in the strong interactions suggests new and very specific relations between the ground-state particles. If a plot is made of hypercharge (Y) against the isotopic spin component (I_3) for all particles of a given spin, they should form a completely regular triangular or hexagonal pattern (Figure 3.2).

The simplest hexagonal pattern (Figure 3.2 *a*) has eight places (two in the centre) and is called an "octet". Gell-Mann and Ne'eman stressed the fact that the known baryons of spin-½ form a natural octet (Figure 3.3) but at that time only seven mesons of spin-0 were known (Table 3.1). Salam and Ward had earlier followed very closely the analogy with charge conservation and predicted yet another octet of mesons of spin 1, analogous to the photon. Since none of these particles were then known, their courage failed them and the manuscript of their paper was kept in a drawer for 6 months, before they submitted it for publication in

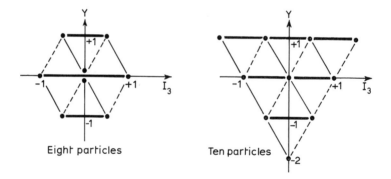

Eight particles   Ten particles

Figure 3.2 *According to unitary symmetry, if particles of the same spin are plotted according to two of their properties – component of isospin (I_3) and hypercharge (Y) – they may form one of two patterns: a regular hexagon of eight particles (with two at the centre), or a regular triangle of 10 particles. The heavy solid lines indicate well-established relationships between particles; the thin and dotted lines show new relationships predicted by unitary symmetry.*

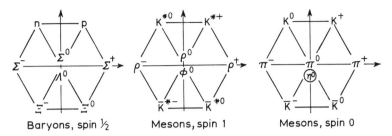

Figure 3.3 *How different groups of particles fit the regular hexagon. The discovery of the eta-zero (ringed) in 1962 completed the set of eight mesons with spin-0, but physicists still remained unconvinced by unitary symmetry.*

March 1961. At the Conference at Aix-en-Provence in September of that year, the discovery of eight spin-1 mesons was announced. They fitted exactly to the predictions. Some evidence was also presented at the same conference for another meson, the eta-zero. Its properties were complicated and elusive, but within 6 months it had been established as the missing member of the spin-0 octet (at the right of Figure 3.3).

In spite of these successes, these particles were all somehow ordinary, and many physicists felt that the whole business might still be illusory. The crux comes with the baryon states of spin $\frac{3}{2}$. By the time of the CERN Conference in 1962, the existence of nine of these had been established. The I_3-Y plot is as shown in Figure 3.4. Gell-Mann was quick to point out that this looked like a ten-particle triangle (Figure 3.2 *b*), but with one particle missing.

The predicted missing particle, called omega-minus, has quite extraordinary properties. It is negatively charged; it is a singlet ($I_3=0$); it has hypercharge of -2 and spin $\frac{3}{2}$.

Even its mass can be specified. Gell-Mann has developed a formula for the mass differences in the hexagon pattern, which agrees with observation, and S. Okubo has shown more generally that the masses of the different isotopic submultiplets of the triangular pattern should be equally spaced. This leads to a prediction of 1685 for the mass of the omega-minus. With this mass, the particle is stable for strong interactions and should therefore have a mean life of about 10^{-10} s.

According to the conservation of baryons and hypercharge, the omega-minus should be produced in collisions between K^- mesons and protons and should decay (weakly, with a change of

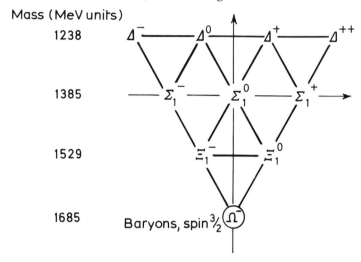

Figure 3.4   *Plotting the nine baryons of spin ³/₂ known in 1963 suggests that one particle is missing – the omega-minus (ringed, at lower vertex). Its properties are very unusual, but can be predicted from unitary symmetry. Does it exist? Read on.*

hypercharge) to a xi-particle and pion, or to a lambda-particle and a K^-.

This is the type of situation about which physicists dream – a startling new theory at an absolutely fundamental level, which coordinates and clarifies a previously confused experimental situation; which is in agreement with the known facts; and which makes an absolutely precise prediction, specific to the new theory and appearing quite weird on the basis of previous ideas.

The technical requirements for the production of the omega-minus are very difficult, and the necessary high energy beams of K^- mesons were only recently engineered at both the Brookhaven and CERN laboratories. Both labs have recently run experiments with large hydrogen bubble chambers. The photographs are being analysed and an announcement may be expected any day.

The confirmation of this "unitary symmetry" in strong interactions will be a dramatic advance in our understanding of the laws of nature. High energy physicists are walking around with a slightly hysterical look, as though they are actually witnessing the apple landing on Newton's head!

20 February 1964

The discovery of the omega-minus particle

The announcement of the discovery at the Brookhaven National Laboratory on Long Island, New York, of the elementary particle known as the omega-minus (as foreshadowed in the article by Paul Matthews) is one of the supreme examples of how to convince sceptics of the validity of a new theory. Nothing succeeds so well as the verification of a prediction based on the theory – particularly if the prediction is something as unusual and unexpected as the omega-minus particle. Its existence was postulated by Murray Gell-Mann of the California Institute of Technology in 1962, on the basis of the "unitary symmetry" theory.

The discovery, by a team of 33 physicists, using the alternating-gradient proton synchroton at Brookhaven, effectively clinches this unconventional hypothesis as to how the particles fit together into "families" with an elegant symmetrical pattern, according to their physical properties. From now on experimenters will be equipped with a powerful means of predicting the properties of undiscovered particles and so constructing further families until they have something analogous, on the nuclear level, to the chemists' Periodic Table.

The idea of unitary symmetry had by no means received the universal support of physicists. In London, it was developed by Abdus Salam with John Ward and Yuval Ne'eman at Imperial College, following a suggestion by the Japanese theoretician Y. Ohnuki. The problem was also taken up by Gell-Mann.

The photograph which has provoked – quite justifiably – the current excitement among physicists is reproduced in Figure 3.5, where the track made by the omega-minus in Brookhaven's 80-inch hydrogen bubble chamber appears as the very short line, before the particle decays into something else after 0.7×10^{-10} seconds. To get this photograph the researchers had to take 100 000 pictures of nuclear interactions.

Not all of the proton synchroton's 33-GeV energy was employed, for the theory predicts that omega-minus particles are most likely to be formed from the interaction between K-mesons and the protons of the liquid hydrogen in the bubble chamber at energies around 5 GeV. The K-mesons in the experiment were produced by bombarding a tungsten target with protons from the synchroton and unwanted particles were then "filtered" out of the beam by electrostatic separators. Analysing magnets removed the

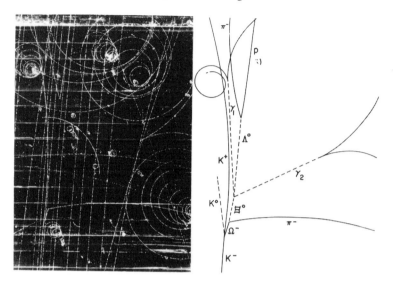

Figure 3.5   *The first observation of the omega-minus (Ω⁻), in the 200-cm (80-inch) liquid hydrogen bubble chamber at the Brookhaven National Laboratory. The particle is produced in the interaction of a negative kaon (K⁻) with a proton (a hydrogen nucleus in the bubble chamber); the paths of the neutrals are inferred from the charged tracks, and an application of the laws of conservation and momentum. (Credit: BNL)*

K-mesons with energies bigger or smaller than 5 GeV. The equipment necessary to send the focused K-meson beam into the bubble chamber was over 100 metres long, and delivered about ten negative K-mesons to the bubble chamber every 2½ seconds. So far the photographs have been only partially analysed. While they may well contain other examples of the omega-minus and its patterns of decay, the one shown here is the only one so far found on which it is possible to check the physical properties.

One of the most startling features of the discovery is the high degree of accuracy with which the particle's properties were predicted. It should have a mass of about 1685 in units of MeV, a lifetime of about 10^{-10} seconds, charge of -1, "spin" of $\frac{3}{2}$, "isotopic spin" of zero, and "hypercharge" of -2. The computed mass of the particle that made the track above is 1686 ± 12 MeV and its other characteristics are as expected. It fits exactly where it should at the apex of a symmetrical triangle that relates it, on the unitary symmetry theory, to nine other heavy particles with spins of $\frac{3}{2}$. The discovery of the omega-minus confirms that other groupings of particles into families or "supermultiplets" are valid. There are still some loose ends in the shape of particles that do not form complete families, and undoubtedly as the energy of the big accelerators goes up and up, more and more families will appear.

27 February 1964

PART TWO

The hunting of the quark

The name of Murray Gell-Mann runs as a constant thread through the history of particle physics in the 1950s and 1960s. Often associated with others, it nevertheless crops up again and again. In 1953, he (and independently, Kazuhito Nishijima) had started to classify the interactions of the newly discovered particles, using a new property called "strangeness". This classification allowed him to predict the existence of the neutral xi, or Ξ^0, 6 years before its discovery. Then in the early 1960s, Gell-Mann (and independently Yuval Ne'eman) took the mathematical symmetries of group theory, and interwove them with the patterns in the observed particles. The strength of this "SU(3)" scheme (SU(3) is the name of the appropriate mathematical group) was revealed in 1964 with the discovery of the omega-minus particle, which had been predicted by Gell-Mann's symmetries. But the SU(3) scheme did not explain why the particles should fit the patterns of this particular group. Not surprisingly, Gell-Mann had the answer; and again someone else, this time George Zweig, came to the same conclusions independently.

Gell-Mann and Zweig noted that the observed particles correspond to higher order – more complex – patterns of SU(3). The simplest pattern, on the other hand, involves the transformations (or interactions) between only three basic objects, apparently unobserved in nature. But the more complex patterns can be derived from the basic triplet, so Gell-Mann and Zweig proposed that there must exist three fundamental entities from which the observed particles are built. In other words, there are new "elementary particles" which combine together to form protons, neutrons, pions, strange particles and so on. Gell-Mann called them "quarks"; Zweig called them "aces". It is Gell-Mann's choice that has remained with us.

Gell-Mann was rewarded for his many contributions to particle physics with the Nobel prize in 1969, as Chapter 4 documents. But the concept of quarks was slow to gain acceptance among physicists. Chapter 4 also includes an account of the quark model, written in 1964, the year the proposals were first aired. It notes one peculiar property of the quarks: they should carry electric charges that are fractions of the observed basic unit of charge – the magnitude of the charge on an electron, e. Two of the quarks are assigned a charge of $-\frac{1}{3}e$; the other a charge of $+\frac{2}{3}e$. This is necessary in order that they can be combined to endow protons and neutrons, for example, with the charges we observe. (In a similar way, one of the quarks with charge $-\frac{1}{3}e$ must carry the property of strangeness, so that it can contribute to the construction of the strange particles.)

This unusual characteristic suggests at first sight that quarks should be easy to spot. Many particle detectors (as Part Seven reveals) work on the basis of registering the ionisation that a charged particle produces as it passes through a medium. This is the process in which electrons are knocked out of atoms in the medium, picking up energy from the passing particle, which consequently slows down. The amount of ionisation is proportional to the square of the charge; so a quark of charge $\frac{1}{3}(\frac{2}{3})e$ will produce only $\frac{1}{9}(\frac{4}{9})$ the ionisation of a similar particle with charge e.

The difficulty that confronted the quark hypothesis in the late 1960s was that nobody could find any evidence for particles with such low ionisation. Nobody, that is, except Charles McCusker. Chapter 5 illustrates how his claim to have observed a quark tracking through a cloud chamber was rebuffed, supported and then again rebuffed. His observation seems certainly to have failed the test of time, in that nobody has ever found similar evidence. Yet McCusker remains convinced that a quark is indeed what he found back in 1969. Such is the way that science works.

The difficulties in detecting quarks, and the lengths to which determined experimenters will go, become still more apparent in Chapter 6, which tells the tale of an experiment by William Fairbank and his team at Stanford University. On several occasions since 1977, Fairbank claims to have measured charges of $\frac{1}{3}e$ on minute superconducting niobium balls, suspended in "mid-air" by magnetic fields. As some of the reports included in Chapter 6 reveal, other teams performing related experiments have failed to find any similar evidence. So Fairbank's results remain a puzzle; it is difficult to believe his results, yet it is even more difficult to find fault with his experiment. What is really needed is for someone to repeat his work exactly.

The quark has proved elusive, slipping easily through its hunters' fingers. Yet it has become fundamental to our understanding of the nature of matter, as I indicated in the Introduction. How can this be?

The answer lies in a series of experiments that began in the late 1960s. Martin Perl discusses some of the results in Chapter 7. Particularly interesting were the measurements made when a beam of high-energy electrons was fired at a target of protons (as in liquid hydrogen) or of protons and neutrons (as in deuterium). The results indicated first that protons and neutrons are not solid lumps, but are "fuzzy", and secondly that within the fuzziness there lie some hard grains, which were named "partons". It seemed that the proton and neutron have some kind of structure, just as Ernest Rutherford's experiments, in which he fired alpha particles at gold foil, had revealed the atom as being mainly empty but with a hard core, or nucleus.

The challenge for the early 1970s was to discover if the partons – the grains in the proton – were indeed the quarks proposed by Gell-Mann and Zweig. Further studies began slowly to suggest that this was indeed the case, as Chapter 7 also reports. To bring the story right up to date, this chapter also includes an account of some of the latest evidence for structure within the proton: results from what in 1982 were the highest-energy man-made collisions ever, between beams of protons and antiprotons.

The real impetus for the quark model came in the period 1974–6, ironically with the discovery of new particles which required new quarks, beyond those proposed by Gell-Mann and Zweig. But that story belongs to Part Three. A final word on the old quarks comes from Harry Lipkin in Chapter 8. He discusses the work of Andrei Sakharov, subjected by the Soviet authorities to internal exile in 1980. Sakharov was one of the first theorists to take quarks seriously and for this reason alone he deserves mention; as Lipkin reveals, he has not lost this interest.

4

The quark hypothesis

Two theorists dare to suggest that protons and their relations are composed of smaller, fundamental particles.

Buried among the many incomprehensible pages of *Finnegan's Wake* what sounds like the pronouncement of an obscure Celtic award runs: "Three quarks for Muster Mark." Appropriately, for modern nuclear physics is often as unintelligible as Joyce to ordinary mortals, the term "quarks" was recently borrowed by Murray Gell-Mann, of the California Institute of Technology, to describe some hypothetical new elementary particles. An experiment performed with the big proton synchrotron at Brookhaven now indicates that quarks do not exist. Or, to be precise, if they do exist, they must have very large masses, by the usual standards of particles.

Gell-Mann, of course, is one of the originators of the so-called unitary symmetry theory which has recently come into prominence as a result of its success in predicting the existence of the omega-minus particle. According to the theory, all the known "strongly interacting" elementary particles and their associated "resonances" can be arranged into families in a way that interrelates their properties. There is one family of ten which includes the omega-minus; this is consistent with the particular kind of algebra which is used, which also suggests that the fundamental family of particles from which all the others can be built up should contain just three members.

None of the known particles can be made to fit the bill and this discrepancy led Gell-Mann to postulate his quarks. Independently, George Zweig, also of Caltech but currently working at the European Organisation for Nuclear Research (CERN), has put

Plate 4.1 *Murray Gell-Mann receives his Nobel prize for physics, from King Gustaf Adolf. (Credit: Keystone)*

forward a similar theory – only he calls his triad of particles "aces".

The really odd thing about quarks or aces is that they are supposed to have fractional electric charges. Ever since Robert Millikan in 1911 did his ingenious oil-drop experiment to measure the size of the charge on the electron it has been part of the physicists' gospel that the electron carries a basic unit of charge which is indivisible. Gell-Mann's three quarks are similar to the proton, neutron and lambda-particle but with charges respectively of $\frac{2}{3}$, $-\frac{1}{3}$ and $-\frac{1}{3}$ of the usual unit charge. At least one of them is expected to be stable so that there should be a chance of detecting it. A particle with a charge of $\frac{1}{3}$ has only one-ninth of the ionising power of an electron so that any tracks it left behind in a bubble chamber would have one-ninth the number of bubbles per centimetre for a given particle momentum; it would emit only one-ninth as much light in a scintillation detector.

L. B. Leipuner, and his co-workers at the Brookhaven National Laboratory, have set out to look for events of this latter kind

(*Physical Review Letters*, vol. 12, no. 15, p. 423). Their statistics show that if any quarks with ordinary masses exist they are extremely rare – less than one for every few thousand million regular charged particles. At CERN, researchers have been scanning previous bubble chamber photographs and an experiment has been running to look for quarks with the 81-cm Saclay–École Polytechnique bubble chamber, again with negative results so far.

The possibility remains that fundamental particles with fractional charges may still exist but they would need to have masses more than three times that of the proton and to generate them would require bigger machines than either the CERN or Brookhaven accelerators.

NOTES AND COMMENTS
23 April 1964

Strangeness, quarks and the eightfold way

Murray Gell-Mann, the 1969 recipient of the Nobel prize for physics, follows in the finest traditions of theoretical prognostication. Gell-Mann, a 40-year-old professor of physics at the California Institute of Technology, was awarded his Nobel prize for "contributions and discoveries concerning the classification of elementary particles and their interactions". The citation is necessarily vague due to the quantity and diversity of ideas that originated with Gell-Mann. He is best known for his recognition and exploitation of the symmetries of strongly interacting particles – which led to his prediction and the subsequent discovery of the omega-minus particle. This work is labelled the "principle of unitary symmetry" or "SU(3) symmetry" or the "eightfold way". Previous to the eightfold way, the number of "elementary" particles had grown to almost 100 but there was no means of unifying them. Gell-Mann devised a classification scheme that incorporated all the known particles, and he unambiguously predicted the properties of new undiscovered particles. This SU(3) model, which was simultaneously and independently derived by Yuval Ne'eman at Imperial College, London, is often described as the particle physics analogue to the chemical periodic table of atomic elements proposed by Dmitri Mendeleev in 1869.

In 1951 Gell-Mann received his PhD from the Massachusetts

Institute of Technology and later joined the faculty at Caltech where he came under the influence of 1965 Nobel Laureate Richard Feynman, one of physics' most inventive theorists. As a joint effort in 1958, Feynman and Gell-Mann proposed "the universal theory of weak interactions" based on "the conserved vector current (CVC) theory" of two Russian physicists, S. S. Gershtein and Ya. B. Zel'dovich. This theory drew an intimate connection between weak interactions and electromagnetic interactions. It is a very rich theory, having led to many detailed predictions.

Moving into strong interaction physics, Gell-Mann examined the then unexplained properties of "strange" particles. He proposed the "strangeness" quantum number and its conservation in strong and electromagnetic interactions. Independently, the Japanese physicist Kazuhito Nishijima struck on the same idea. In 1961 Gell-Mann unfolded the eightfold way, which was based on eight quantum numbers, including strangeness, and named after the eight truths attributed to Buddha. According to SU(3) symmetry, particles with the same intrinsic spin angular momentum and parity can be grouped in multiplets in a definite fashion. Gell-Mann derived a relationship between the masses of the members of one multiplet that was later generalised by the Japanese physicist, S. Okubo. This famous relationship, now called the Gell-Mann–Okubo mass formula, was a big breakthrough to understanding the behaviour of elementary particles.

Gell-Mann's crowning achievement was his prediction of the omega-minus at the apex of a decuplet of heavy particles including delta, sigma and xi particles (see Chapter 3). He accurately anticipated all of its quantum properties including mass, spin, parity, strangeness and electric charge. The first observation of the omega-minus at Brookhaven National Laboratory in early 1964 was a great triumph for Gell-Mann and the theories of elementary particles.

Another famous contribution of Gell-Mann's is the invention of "quarks" as the bases of elementary particles (see last item). Again, this idea was independently derived, this time by George Zweig at CERN. Quarks are named by Gell-Mann from a passage in James Joyce's *Finnegan's Wake*. They have the unique property of fractional electrical charge. Also noteworthy is Gell-Mann's elucidation in collaboration with Abraham Pais of the dynamics of neutral K mesons and the role of isotopic spin in weak decays.

As can be seen, theorists do not generally work in a vacuum. It is difficult to point at a physical theory and say "this is uniquely Gell-Mann". However, it can be said that Gell-Mann has had a finger in every important theoretical pie over the past 15 years. His creativity must have been directed in some way by the aphorism of Buddha that led to the name eightfold way. "Now this, O monks, is noble truth that leads to the cessation of pain, this is the noble *Eightfold Way*: namely right views, right intention, right speech, right action, right living, right effort, right mindfulness, right concentration."

6 November 1969

5

Quark tracks down under

A quark leaves a track in a detector in Australia. Or does it?

Charles McCusker of the University of Sydney created quite a stir at the 11th International Conference on Cosmic Rays in Budapest by announcing that he had detected cosmic ray particles of fractional electric charge – a property of the "quark" which has been suggested as being the building block of elementary particles. But a critical analysis of his evidence by other experts at the conference casts doubts on McCusker's claims. From 60 000 tracks photographed in four Wilson cloud chambers at sea level, five tracks exhibited characteristics of fractionally charged particles, but these tracks may be spurious events of known elementary particles.

McCusker used an array of Geiger counters and scintillators to activate the cloud chambers for cosmic rays of 4×10^{15} eV energy. The charge on the particles was determined by measuring the density of the droplets formed by primary ionisation of cosmic particles in the chamber. Since the droplet density is proportional to the square of the charge, particles with $2/3$ electric charge (a predicted property of quarks) would produce $4/9$ of the primary ionisations caused by a particle of unit charge. In the five anomalous tracks, the droplet density was about half of that for the other 60 000.

However, the droplet density can fluctuate from other causes such as variation of the particles' velocity and secondary ionisation effects. Most experts agree that high-energy cosmic rays are the most likely source of quarks, but they think that McCusker did not account for all spurious events, and place the confidence level on his discovery at about 10 per cent.

MONITOR
11 September 1969

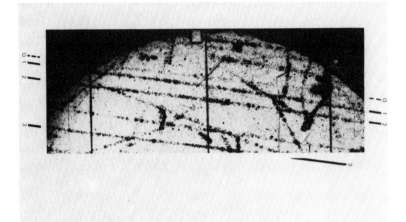

Plate 5.1   *Event number 66240 in a cloud chamber of the Sidney quark search, shows a faint track QQ. This track has fewer drops along its length than have tracks 1, 2 and 3, which are due to fast electrons in a shower produced by cosmic rays high in the atmosphere. The faint track could have been produced by a particle with less ionising power than an electron, such as a quark with fractional charge. (Credit: C.B.A. McCusker)*

Monitor

Last September, Charles McCusker of the University of Sydney, Australia, announced that he had detected quarks in the cores of cosmic-ray air-showers (*Physical Review Letters*, vol. 23, p. 658). If substantiated, this would be a most important discovery as theorists had hypothesised that quarks were the basic building blocks of all matter. McCusker claimed that he and a team of three physicists had observed particles possessing ⅔ the basic electric charge – a predicted property of quarks. At the time, many experts took exception to the Australians' interpretation of their data and few accepted McCusker's conclusions. In a recent publication, R. K. Adair of Yale University, Connecticut, and H. Kasha of Brookhaven National Laboratory, New York, spelled out their specific objections to the Australian result (*Physical Review Letters*, vol. 23, p. 1355).

Adair and Kasha pointed out that no less than eleven other

experiments have been performed with sensitivity comparable to that of the Australian work and none of these showed evidence for quark production by cosmic rays. Using the flux of quarks as calculated from the Sydney observation, the two antagonists reported that they should have detected 1000 quarks in their cosmic-ray experiment; they saw none. However, this does not explain the presence of the anomalous tracks observed by the Australians in their cloud chambers.

Adair and Kasha offered several alternative explanations for the quarks. They claimed that if properly analysed, the quarks could be explained as statistical fluctuations of the interactions of electrons and muons in the cloud chambers. Electrons of low velocity would produce tracks identical with the high-velocity fractionally charged quarks. The electrons could be produced from interactions of the air-shower with materials about the apparatus. The two Americans also noticed that the tracks left by the "quarks" were not parallel with the other cosmic-ray tracks in the chamber. As the air-shower originated at the top of our atmosphere, this small angle between the tracks was of sufficient size to make the "quark" miss the detector array if it were created in the air-shower. This evidence showed that the "quarks" probably did not belong to the core of the air-shower and were most likely secondary electrons.

The onus is now on the Australians to defend their results, but it will be difficult in the face of such overwhelming evidence to the contrary. The hypothetical quark will probably remain hypothetical after the dust has settled.

25 December 1969

Monitor

At a time when the experimental discovery of quarks – hypothesised as the most elementary of elementary particles – is under heavy criticism, new evidence is appearing which may add substance to the claims of Charles McCusker of the University of Sydney. Last September McCusker reported his observation of cloud-chamber tracks left by quarks produced in cosmic-ray air showers. Since then several papers published in *Physical Review Letters* questioned McCusker's conclusion. Now, McCusker has countered some of the criticism, and as added support, a bubble-chamber picture from Argonne National Laboratory, Illinois, contains a track which could be explained by quarks.

McCusker identified his quark by the low density of ions produced along the cloud chamber tracks. Since quarks are supposed to possess either ⅔ or ⅓ of the unit electric charge, they should produce less ionisation than other particles – provided all the particles have the same velocity. This fact, that ionisation in a cloud chamber depends both on electric charge and velocity of the incident particle, has been the main point of McCusker's critics. They say that the low-density tracks might be explained by fluctuations in the velocities of known particles rather than the introduction of quarks. They urged McCusker to analyse carefully all of his 60 000 tracks – of which five are quark candidates – to see if statistical fluctuations could account for the low-density ones (quarks with ⅔ charge should produce 4/9 ionisation compared with known particles). As reported at the Australian Institute of Nuclear Science and Engineering, McCusker has analysed 300 tracks and has concluded that fluctuations in the known particles' velocities cannot account for the quark tracks. There are still other criticisms to answer, but the main objection to McCusker's quark is on the wane. The news from Argonne is also very encouraging to McCusker's cause. During tests of Argonne's large bubble chamber, cosmic-ray tracks were photographed. Physicists from Ohio State University, Rose Polytechnic Institute in Indiana, and the University of Kansas observed a track in one of their photographs which is a likely candidate for a quark. If the particle has ⅔ electric charge, its mass is 6.5 GeV (thousand million electron volts) or less. Although they cautioned about drawing firm conclusions from only one track, the case for quarks is growing stronger.

12 March 1970

Monitor

High-energy physicists seem to be engaged in a new game – "quark baiting". First one experimental group claims to have detected a quark – the hypothetical particle which is supposed to be the building block of other elementary particles – and then another group or groups present masses of evidence designed to show that we are being deceived. The "quarks" detected by Charles McCusker and his colleagues from the University of Sydney have come under heavy fire. Now, the bubble-chamber track identified by W. T. Chu and Young S. Kim, Ohio State University, W. J. Beam, Rose Polytechnic Institute, Indiana, and

Nowham Kwak, University of Kansas, as originating from a cosmic-ray quark, is having its day in court.

Chu and his associates photographed a bubble-chamber track in the Argonne heavy-liquid chamber which they claim was caused by a fractionally charged particle – a predicted property of quarks (*Physical Review Letters*, vol. 24, p. 917). Defending the integrity of their institution and its equipment, five physicists from Argonne National Laboratory have demolished the arguments of Chu and his team (*Physical Review Letters*, vol. 25, p. 550).

If the quark exists, it can be detected in a bubble chamber by the low density of the bubbles left in its trajectory. However, low-density tracks can result from causes other than fractionally charged particles. For example, the bubble density also depends on the particle's velocity.

W. W. M. Allison and his four Argonne colleagues analysed 151 pictures on the same roll of film as the "quark" and studied the spectrum of the bubble densities for the cosmic-ray tracks. They found that several tracks have densities lower than that of the "quark". Thus it can be explained by normal statistical fluctuations. They presented other evidence based on the size of the bubbles which indicates that the "quark" is most likely a muon which arrived 10 milliseconds before the camera flash. They conclude: "The important point here is not whether the quark explanation is possible but whether it is required." Obviously it is not.

3 September 1970

6

Fractional charge: the mystery of the niobium balls

DAVID PATTERSON

American physicists claim to have measured charges that are fractions of the electronic charge, but other experiments have failed to find similar evidence.

Has the quark been found at last? Reaction to the discovery announced recently has been cautious. The discovery would be very important – even industrially important, for quark-catalysed reactions would open up a whole new branch of chemistry. The interpretation of the experiment is complicated, so no-one is rushing to conclusions. But there is a buzz of excitement about for the principal researcher involved, William Fairbank of Stanford University, has a high reputation. He and his colleagues have now published their results in *Physical Review Letters*, vol. 38, p. 1011.

The object is to take a sequence of measurements of the charge on a 0.25mm superconducting niobium ball suspended in a magnetic field. The ball is as large as possible, to sample as much matter as possible for quarks, but not so large that it is impossible to measure a single electronic charge on the ball. The charge on the ball is found by measuring the ball's motion in an electric field. It takes about 80 seconds to make one measurement of charge. After each measurement the ball is subjected to a stream of electrons or positrons from radioactive nuclei in an attempt to discharge it and the charge measured again. In this way a plot of the frequency of various charges on the ball is built up.

If the plot is displaced from zero, as Fairbank's is, there must be a fractional charge on the ball. Or must there? Fractional charges can be mimicked by a number of different small forces to which the ball is subject, and everything depends on eliminating or measuring them.

Fairbank and friends claim to have done that and claim events of charges -0.331 ± 0.070 and $+0.337 \pm 0.009$ in a series of only 11 runs (the other runs are consistent with zero). The events came when the balls had been in association with the heavy element tungsten, and one explanation of why other groups looking for natural quarks have not found them before is that they have never looked at heavy elements. Perhaps quarks like it heavy.

MONITOR
5 May 1977

Queasy over the quark

A small basement room in the physics laboratory at Stanford University may become celebrated as one of those places where there was a turning point in science, or it may turn out merely to be the scene of yet another false alarm that draws the public's attention to science's unsteady progress. Picture a perfectly ordinary laboratory with pieces of apparatus on racks, the occasional poster, a large hole in the wall showing signs of builders' activities and a dustbin-like piece of cryogenic equipment protruding from the floor and you have some idea of the place where graduate student George LaRue and his supervisor Professor William Fairbank detected what they believe to be the signature of some object carrying fractional electric charge. Although neither Fairbank nor LaRue claims that the fractional charge is borne by a quark, there is not much doubt that this notion is uppermost in their minds. LaRue certainly feels that his apparatus has been purged of all misleading influences that could raise false hopes − "if it's not a quark, what is it?"

Since the publication of their paper "Evidence for the existence of fractional charge on matter" in *Physical Review Letters* (vol. 38, p. 1011), LaRue reports that there has been no rush of adverse criticism of the experimental technique. Yet undoubtedly at Stanford, as elsewhere, there is an atmosphere of unbelief, that ranges from "I don't *not* believe it" to "it's a hundred to one that it's not right".

What is certainly not in doubt is the ingenuity of William Fairbank as an experimenter: he has gained a reputation for undertaking exceedingly difficult experiments of a crucial nature in all branches of physics. A check list of his current activities is a testament to the breadth of his interests. In addition to the search for fractional charge, he and his colleagues are engaged on a space-borne gyroscope to test general relativity, a test to find out whether or not positrons and electrons fall oppositely in a gravitational field, a test of the principle of equivalence to be done in space to a precision of one part in 10^{17}, and a search for the elusive gravitational radiation, as well as a host of ingenious devices intended to put advanced physics methods to use in medicine and physiology. The scope of Fairbank's interests is admired by colleagues and indicates a competitive if not aggressive attempt to leave his name in the annals of physics.

Fairbank's striving in the quark hunt is no mere passing interest – he was aroused by the publications of Murray Gell-Mann and George Zweig in the mid-1960s that suggested that quarks might have non-integral electric charge. Soon Fairbank had graduate student Art Hebard working on an apparatus that might detect whether there were in nature particles with fractional charge. Hebard's thesis, published in 1972, opens with a quotation, but not in the grandiose literary style beloved by fundamental scientists. Instead it's drawn from a publication of Robert Millikan – the grand-daddy of searchers for the size of the electric charge carried on particles of matter. Millikan wrote in *The Philosophical Magazine* of 1909 "I have discarded one uncertain and unduplicated observation apparently upon a single charged drop, which gave a value of the charge on the drop some 30 per cent lower than the final value of *e*". One wonders what would have happened to the course of science if Millikan had been able to duplicate his dubious result?

Hebard and Fairbank's apparatus did not provide the unequivocal evidence in 1972 that was needed to convince a sceptical physics community of the existence of fractional charge and Hebard left for the Bell Laboratories in New Jersey. His place in the basement room was taken by George LaRue, a physics graduate of the University of Maryland who had come to Stanford with the aim of working at the Linear Accelerator Center. LaRue changed his mind after seeing the size of the groups doing experiments at SLAC and decided that a table-top experiment suited his temperament better.

The table-top experiment that LaRue inherited was Hebard's apparatus – then an up-to-date version of the Millikan experiment beloved of physics students. Millikan had shown that it was possible to measure the electric charge on an oil droplet moving in an electric field: in essence, he injected a cloud of oil droplets from

Plate 6.1 *Robert Millikan measured the charge on the electron in 1909 by measuring the motion of charged oil drops falling through an electric field. William Fairbank and his colleagues have devised a similar experiment, updated with modern technology, to search for the fractional charges expected for quarks.* (Credit: The Illustrated London News)

an atomiser into a region between two plates, applied an electric potential across them and watched through a telescope how an individual droplet moved. He was able to alter the electric charge on the droplet by moving in a radio-active source and so could measure the rate of movement with different levels of charge. The Stanford experiment brought in the sophistication of contemporary cryogenic technology and a host of complex electronics to refine the basic Millikan experiment.

The key feature not accessible to Millikan in 1909 was that of superconductivity. Fairbank, Hebard and LaRue used a niobium sphere instead of an oil droplet as the charge carrier; niobium is superconducting at liquid helium temperatures. The ball can be made to float in a vacuum by passing a current through super-conducting niobium coils beneath the ball. Once the current has been induced in the coils, it continues to flow and maintain the ball in a steady state of levitation. The balls can thus be larger than Millikan's oil droplets – his droplets had a mass of about 3×10^{-11} grams, whereas the niobium spheres were 0.25 mm diameter and had a mass of roughly 10^{-4} grams.

One of the ideas underlying this approach to finding fractionally charged particles is the notion that there should be a few free quarks attached to otherwise normal matter. This is not so unlikely; for quarks may have been made shortly after the big bang and all of them may not have combined to form integrally charged objects. Also it has been argued that because cosmic rays have been bombarding the Earth (and before that the pre-solar nebula) since time began, some of these collisions may have had enough energy to disrupt the nucleons and allow the quarks to come out and associate themselves singly with stable atoms. There's no doubt that quarks are rare, so the chances that two quark-doped atoms might "get it together" and annihilate their fractional charges are negligible. So, the larger the sample of matter the greater the chance that a quark might be trapped. Hence the niobium balls instead of the lightweight oil droplet.

The next step is to apply an electric field to the ball and see if it moves in the manner appropriate for a ball carrying a fractional charge. The technique adopted by Fairbank and his students was characteristically ingenious.

The group levitated the ball between two capacitor plates roughly one centimetre apart (Figure 6.1): in the absence of an electric field, the levitated ball behaved like a massless magnetic spring and bounced up and down roughly once a second. Fairbank and colleagues then used an *alternating* electric field to move the ball. The apparatus applied to the ball a 2000-volt peak-to-peak square wave at the ball's up-down resonant frequency. Every 50 cycles, the phase of the voltage was reversed. The reasons for this sophistication are many – the resonant technique boosts the sensitivity of the method dramatically, the square wave voltage shape allows more energy to be added to the ball per cycle than, say, a sine wave, and so speeds up the measurement process, and the reversal of phase cancels out some of the inhomogeneities of the field.

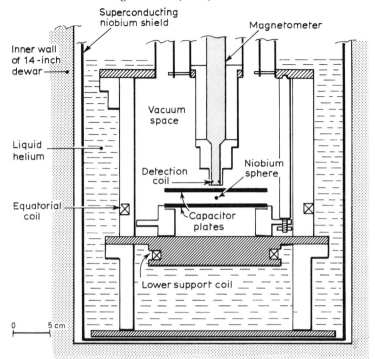

Figure 6.1   *A charged niobium ball floats on a magnetic cushion. An electric field nudges the ball; how much the ball moves determines its charge.*

Once the ball is suspended in the magnetic field, and then made to move under the applied electric field, the next stage is to measure how the ball responds – in this case, how it picks up energy from the electric field and increases its amplitude of oscillation. Again superconductivity comes to the rescue – in the form of a superconducting loop closely coupled to the magnetisation of the ball and in series with a second remote superconducting coil. The latter couples to a Superconducting Quantum Interference Device (a SQUID) which acts as a magnetometer and generates a measurable output voltage – allowing displacements of the ball between the plates as small as a micrometre to be measured. The SQUID magnetometer was chosen because it provided a way of detecting tiny flux changes. The system is activated by the coupling of the niobium ball's inevitable trapped magnetic field with the primary coil as the ball moves up and down under the influence of the applied electric field.

The biggest problem is that there are many extraneous electric forces that could act to mimic a charge on the ball. Hebard writes in his 1972 thesis of an earlier, unpublished "quark": "Preliminary results obtained for a single sphere indicate a non-zero residual charge of magnitude -0.37 ± 0.03. This value of approximate magnitude $-\frac{1}{3}$ does not necessarily imply the reality of quarks because of the fact that there might exist spurious charge forces which are caused by the apparatus. Such a force might arise for example if the induced electric dipole moment on the sphere, which is proportional to the electric field, interacts with a fixed field gradient, caused by a surface dipole layer on the capacitor plates, to give a force linear in the electric field and hence indistinguishable from the charge force." Fairbank has now been able either to eliminate these spurious forces or to reduce them to a level where they would count only for a fraction of a per cent of the charge on an electron.

LaRue seemed placed to make a dramatic announcement last September, but he waited because another of William Fairbank's colleagues had made a discovery that was to prove invaluable to LaRue. One of the spurious forces in LaRue's apparatus was the "patch effect" – a phenomenon that allows the electric potential on a plate to be unevenly distributed over its surface. To try to avoid the uncertainties that would derive from these voltage patches LaRue had to measure the ball's movements in various positions between the titanium capacitor plates and hope to calibrate them out. But while doing the measurements on electron and positron fall in a gravitational field, J. M. Lockhart had discovered that the patch effect disappeared in copper, when the metal was cooled down to a temperature of 4.2K. LaRue saw the advantages that this phenomenon held for his apparatus. By the beginning of this year, he had built copper plates into his apparatus.

You might think, given such a sophisticated level of instrumentation, that thereafter the search for fractional charges was simple – something that could be done in a matter of minutes or hours. Yet LaRue found that at best it would take days – sometimes a week – to obtain all the results for each ball. The experiment in practice turned out to be somewhat socially unacceptable – each run had to be done at night time when the rumble of trucks and other ground pounding activity had dropped to a minimum.

Given that so much night work had to be devoted to measurement, it must have come as quite a blow to LaRue that one ball which seemed to bear fractional charge lost it when it was taken out of the apparatus. This ball, between runs, was treated in the normal way – cleaned in alcohol and acetone and stored in a bottle – and then put back between the plates. After one such treatment, it behaved as if it no longer carried fractional charge. But even in adversity, one idea was derived from the loss – the fractional charges, if such there be, are perhaps carried on the surface of the balls and are thereby vulnerable.

Both LaRue and Hebard had found that other aspects of surface treatment of the balls greatly affected their behaviour. If for example a ball was work hardened before being tested, its oscillations were damped to an undesirable degree. Consequently each ball was carefully annealed before testing. Some of the balls were heat treated on a niobium substrate and some on a tungsten substrate. After the heat treatment, the balls are placed on a small plastic carrier that allows the ball to be swung into a roughly central position between the plates.

The whole apparatus is then cooled to below 4.2K and the ball is floated off the plastic carrier by raising the current in the levitating coils beneath the ball. The ball is guided accurately to the mid-point between the plates – an observation made through a telescope mounted above the 2-metre-tall apparatus – by adjusting the current in the levitation coils. The time taken to float the ball properly may be several hours. The heat exchange gas, used for cooling, must be pumped out to produce a vacuum of 10^{-4} torr to keep down the friction on the ball. After thermal equilibrium has been achieved – again a period of several hours – the bulk of electric charge on the ball has to be neutralised. It's not untypical to find that the ball carries 100 000 charges on insertion and bringing this down to the level of a few electron charges takes considerable time. Sodium-22 and thallium-206 are the sources of positrons and electrons used to add or subtract charge and capsules of these elements can be moved close to the ball during the course of an experimental run. Equally, they can be shielded when measurements are being taken.

Once this state of the apparatus is reached LaRue is then set to measure the way that the ball behaves when the alternating electric field is applied. The key measurement is concerned with the speed at which the ball picks up energy from the electric field. LaRue measures the increase in the amplitude of the oscillation during the period when the charge on the ball is constant – typically 10 to 15 minutes. One of the complicating factors is the capture of charge by the ball from background radiation like cosmic rays – after a week or so in the apparatus this generally occurred at the rate of between three and six electron charges per hour. Despite all the difficulties, LaRue is convinced that he is able to resolve charges as small as 1 to 2 per cent of the charge on the electron.

In his most recent series of tests, LaRue used a set of eight balls. Five were conventionally heat-treated on a niobium substrate and the other three on a tungsten substrate. Two of the three balls heat treated on tungsten showed $\frac{1}{3}$-integral electron charges: the third of the tungsten treated balls showed a net zero charge within the limits of the experimental accuracy.

In their paper, Fairbank and LaRue conclude "we report evidence for fractional charges close to $\pm\frac{1}{3}e$ apparently transferred to niobium from a tungsten substrate which we cannot explain by magnetic or multipole forces".

Nevertheless one wonders, if the results are to be taken at face value, what is it about tungsten – specifically about *that particular* piece of heat treating substrate – that conveyed fractional charge to two balls? It cannot simply be some property of tungsten in general since the niobium from which the balls are made contains both tungsten and tantalum as impurity. Clearly tungsten is not a quark-catcher as some theorists wanted to believe after the meeting at which the results were first announced. Fairbank's observation was that there might be something special about the history of that particular piece of tungsten that gave it these special qualities of inducing fractional charge on two balls.

Of course the matter does not rest there. Fairbank is now mustering an army of students to take up quark searches 24 hours a day over a long period so that some kind of abundance statistics can be built up. Also spurred on by this triumph, he and LaRue are taking further guidance from the history of physics for yet another experiment – an e/m experiment – in which they hope to measure the charge to mass ratio of the fractionally charged objects they have discovered. LaRue hopes to have this experiment under way within the year. The scheme would involve vaporising a ball carrying a fractional charge and applying an electric field that would separate the charged from the uncharged particles. LaRue hopes that time-of-flight detectors should allow them to find the mass of the fractionally charged particle.

A clear-cut confirmation of the existence of a free quark would be a real headache for mainstream particle theorists. The claim for finding a fractionally charged particle and the possibility of finding its mass would be heady stuff for even the most hardened experimenter – yet LaRue (even if he is excited) doesn't really show it. The nearest he came to waxing prolix about his work during our conversation was "I guess I'm excited – it's kind of thrilling because no one's seen anything like it before. You've just got to hope that you're right."

30 June 1977

Monitor

Excitement following the apparent discovery of free quarks by Stanford physicists LaRue, Fairbank and Hebard may be dampened by the failure of another very sensitive experiment to find any evidence for them. In a recent edition of *Physical Review Letters* (vol. 38, p. 1255) a second trio of experimenters, G. Gallinaro, M. Marinelli and G. Morpurgo from the University of Genoa, report the continuation of an experiment they began in 1966 – and their conclusion is that there is less than one quark for every 3×10^{21} neutrons and protons in the sample of iron which they studied. Not only that, but a further experiment, announced last week, has looked specifically at the element Fairbank supposes the free quarks to reside on – tungsten – and found no evidence for them.

The experiment of the Genoese, like that of the Americans, is an up-to-date version of Millikan's experiment: the sample whose charge is to be measured is suspended in a vacuum between the plates of a capacitor, and when an alternating voltage is applied the test sample oscillates in the resulting electric field with an amplitude proportional to its charge. The Italian physicists performed their experiment on small pieces of iron weighing around two thousandths of a gram. The samples were suspended in a magnetic field regulated by a feedback system to maintain their vertical position in the horizontal electric field.

The experimenters used cylindrical samples and spun them about a vertical axis at 30 to 40 rotations a second. This spinning washed out troublesome electric dipole interactions which are known to mimic fractional charges. After detailed examination of possible sources of error, Gallinaro, Marinelli and Morpurgo feel confident in reporting neutrality of the three cylinders so far tested to within one-tenth of the charge on the electron. They are now planning to improve their apparatus to hold larger samples.

There thus seems to be a contradiction between the Stanford and Genoa experiments. But this may only be apparent. The environmental predilections of quarks, if indeed they exist, remain to be established. The negative outcome of the Genoa experiment strictly applies only to the particular sample of iron from which the test cylinders were made. In the Stanford experiment super-conducting niobium spheres were used, and of three of these spheres which had been in contact with a tungsten substrate during heat-treatment two were found to have fractional charges, which has led Fairbank to speculate that there may be something "special" about his tungsten.

This led the other experimenters – working on a much simpler modification of Millikan's technique – to look at tiny specks of tungsten released from a tungsten arc. If tiny specks of tungsten on the niobium ball contain free quarks, then why not look at tiny specks of tungsten?

So a group of physicists from San Francisco State University measured the electric field strength necessary to suspend a tungsten speck against gravity (*Physical Review Letters*, vol. 39, p. 369). They then charged and discharged the speck with positive and negative β-radiation (each particle of which – an electron or a positron – carries unit charge). Each time the speck changed charge the physicists measured the electric field required to suspend it. When the speck had its minimum negative (or positive) charge, a minimum positive (or negative) field was required to suspend it against its own weight. If there was no fractional charge on the particle these two field strengths had to be equal. They were, for hundreds of measurements on 69 particles. The physicists seem to have found that if fractional charges exist in tungsten they must come at a concentration less than one in 10^{12} atoms of the element.

25 August 1977

Monitor

After nearly two more years of painstaking work, William Fairbank and his colleagues at Stanford University have caught further glimpses of one of the faces of the elusive quark. In 1977 Fairbank started a controversy by reporting that he had measured a charge of only $\frac{1}{3}$ that of the electron, residing on $\frac{1}{4}$mm diameter niobium balls. And if quarks, the much desired but unseen hypothetical building blocks of protons and neutrons, do exist, then they should carry "fractional charge".

The new results, reported in *Physical Review Letters*, vol. 42, p. 142, are from a series of measurements made with modified apparatus. New techniques allow the team to reduce the errors due to background forces to less than 1 per cent of the force from the applied electric field.

As before, Fairbank and friends find charges of $\frac{1}{3}e$ (where e is the charge of the electron) on some of their nine niobium balls. What is more remarkable, however, is that the charge can be made to disappear, and then to reappear again, by applying an electric discharge. The discharge seems to result from the emission of electrons from a ball as the electric field between the capacitor plates is increased. Even more surprisingly, the $\frac{1}{3}e$ charge can be removed by washing the balls in acetone and alcohol.

Fairbank interprets these results as an indication that the charge resides on the surface of the balls. Further evidence for this picture is given by the crystalline structure of the balls. Initially they are multi-crystalline, but after the heat treatment (which all the balls undergo before the measurements are made), they become single crystals. It may be that processes occurring during the heat treatment carry the $\frac{1}{3}e$ charges to the surface.

So the question as to whether free quarks have been found has once again arisen. And the theoreticians who are busily trying to prove that quarks cannot be free must be feeling slightly uncomfortable. However, it must be remembered that the observation of $\frac{1}{3}e$ charge is not sufficient to prove that quarks exist, but if confirmed, it would be in itself a tremendous discovery.

1 February 1979

Monitor

Physicists at the Ohio State University have been investigating the claim that matter can carry electric charges which are fractions of the charge on an electron. The Ohio team has followed up work by William Fairbank and colleagues at Stanford University, who found evidence for charges of $\frac{1}{3}$ of e, the electronic charge, residing on the surface of minuscule balls of niobium. But the new research has revealed no trace at all of fractional charges.

The researchers at Ohio looked for fractional charges in helium gas, because, as they point out, Fairbank did his experiments in helium. So the fractional charges may originate in the helium as opposed to the niobium spheres themselves.

The experimenters used a 1.5 MV Van de Graaff accelerator to produce an energetic beam of helium ions which they then passed through successive magnetic and electric fields to separate the particles in the beam according to their mass-to-charge ratios. As the accelerator gave all particles of the same charge the same energy, a detector which then measured their energy also indicated their charge (*Physical Review Letters*, vol. 43, p. 1288).

From Fairbank's results, the Ohio team expected to find nine fractional charges per 10^{14} helium atoms. Instead they found that such charges could exist only at the level of less than one for every 3.9×10^{14} helium atoms. The researchers looked particularly for charges of $\frac{1}{3}e$ and $\frac{2}{3}e$, in the ranges of 0.13 to 7.5 and 0.27 to 15.0 atomic mass units (1 amu is $\frac{1}{12}$ the mass of a carbon atom). The experiment would have detected fractional charges either attached to helium atoms or nuclei, or on free quarks – the postulated constituents of all sub-atomic particles.

29 November 1979

Monitor

They've done it again. Scientists at Stanford University, California, have published more results indicating the existence of charges of one-third that of the electron, on minute balls of niobium. Out of 21 new measurements, they have found four charges of $+\frac{1}{3}e$ and four of $-\frac{1}{3}e$ where e is the electronic charge (*Physical Review Letters*, vol. 46, p. 967).

William Fairbank and his colleagues at Stanford first reported evidence of charges of $\frac{1}{3}e$ in 1977, and again in 1979, and since then they claim to have further modified and improved their technique.

The researchers have performed the experiment a total of 40 times, using 13 different balls, with radius 98, 116 and 140 micrometres. Five of the balls have exhibited fractional charge on a total of 14 occasions.

Fairbank, George LaRue and James Phillips, now believe they have fully taken into account all background forces that are not negligible, and say they are "forced to conclude that fractional charges with magnitude $\frac{1}{3}e$ must exist". The researchers also find that the charge on a ball can change by $\frac{1}{3}e$ – that is, between 0 and $\pm\frac{1}{3}e$, or vice versa. These changes occur only between levitations, when the balls have come into contact with some other surface.

The experiments at Stanford are being closely watched by particle physicists, who believe the ultimate constituents of nuclear matter to be quarks – entities that carry charges of $\frac{1}{3}e$ and $\frac{2}{3}e$. But individual quarks have never shown up in experiments at particle accelerators, with cosmic rays, or in more exotic locations such as moon rocks and oysters. Indeed, the currently popular theory of the dynamics of quarks suggests that they cannot appear alone, but must always be tied to other quarks or anti-quarks within the observed sub-atomic particles.

In an experiment somewhat similar to that of Fairbank's, performed at the University of Genoa, M. Marinelli and G. Morpurgo last year reported that they had found no evidence for fractional charges on a 3.7-milligram sample of steel (*Physics Letters*, vol. 94B, p. 427, and p. 433). They put lower limits on the existence of fractional charge than the experiment at Stanford implies.

30 April 1981

Monitor

A new experiment based on the technique Robert Millikan used for measuring the charge on the electron in 1911 has found no evidence for electric charges that are fractions of that on an electron. This adds weight to other searches that have failed in the same quest, while an experiment at Stanford University, by William Fairbank and collaborators, remains the only one to claim to have discovered fractional charge.

In his original experiment, Millikan allowed drops of oil to fall between metal plates. By measuring the electric field required to balance the gravitational force on the droplets, Millikan could calculate their electric charge. His results suggested that charge comes only in multiples of e, the elementary charge on the electron. But more recent experiments in particle physics suggest that protons, neutrons and other particles are made up of more elementary constituents – quarks – which have the unusual property of possessing charges that are only fractions of e.

In the latest experiment, Christopher Hodges and colleagues at San Francisco State University have measured the charge on some

100 000 small drops of mercury, ranging in size from 3.5 – 6.5μm (*Physical Review Letters*, vol. 47, p. 1651). The researchers illuminated the drops with light from an argon laser, and observed the images as the drops fell past a grid of horizontal slits, through regions of differing electric field. By measuring the time taken to pass from slit to slit, they could calculate the velocity of each drop in each region, and hence the drop's electric charge. In a total of 175 μg of mercury – 60 μg refined, the rest "native" mercury from the Socrates mine in California – the team found no evidence for charges of $\pm$ $\frac{1}{3}e$, the charge expected for one type of quark present in protons and neutrons. That is equivalent to there being less than one free quark for some 10^{21} quarks bound in the protons and neutrons in the mercury – in line with other experiments, except, of course, the one at Stanford.

14 January 1982

7

The grains in the proton

MARTIN PERL

Quarks slowly reveal themselves, not as individual particles, but as structure within the proton.

We think of the elementary particles – the protons, mesons, electrons, photons, and so forth – as very small and fuzzy billiard balls. They are too small and too fuzzy to measure or to cut apart: we can only shoot one ball at the other and if they collide we can see what happens. Physicists often use this analogy, but we must remember that these elementary particle billiard balls have some very strange properties. Thus if an accelerator shoots a pi-meson at a stationary proton, sometimes the pi-meson just bounces off the proton. The pi-meson and proton then recoil in different directions just as in billiards. But sometimes two completely different particles such as a K-meson and a lambda can come out of the collision. It is as though the collision of two billiard balls produced a golf ball and a croquet ball. If there is enough energy then many balls, all different in some cases, are created.

One of the simplest and most fundamental questions that one can ask about the elementary particles is what is the overall collision or interaction probability between a pair of particles? This property, called the total cross-section, is a measure not only of the particles' size but also of the strength of the forces between the particles. Two particles need not touch to interact: they can attract or repel each other at a distance. Actually, because the particles are very fuzzy, there is no sharp distinction between actual contact and action-at-a-distance. At low energies the total cross-section depends strongly on the energy and fluctuates in some cases quite drastically with energy. It is as though the size of the billiard balls depended upon their speed or energy. At high energies up to 100 GeV, the total cross-sections for some pairs of particles become constant and independent of energy. The elementary particle physicist had in a way always hoped to reach an energy region where total cross-sections would become independent of energy because he thinks that the theory may be simpler here. But having reached that energy for some pairs of particles (the negative pi-meson and the proton, for example) but not for other pairs of particles (anti-proton and proton, for example) the physicist is not sure what that theory should be. In fact, some theorists merely use a slightly sophisticated version of the billiard ball picture, regarding the radius of the billiard ball as an energy-dependent parameter that becomes constant at high energy. Other theorists suppose that the radius increases with energy. But these theorists also regard the elementary particles as being partially transparent to each other so that sometimes the particles can pass through each other without interacting. To obtain a constant total cross-section at high energy, these theorists must then assume that as the energy increases the transparency increases: the increasing transparency just compensating for the increasing radius.

Another basic question concerns the internal structure of elementary particles such as the pi-meson or proton. Is the matter inside these particles homogeneous or are there yet simpler objects inside these particles? The physicist would like to find such simpler objects because there are now so many different elementary particles that the name is almost a misnomer. Perhaps these different elementary particles could be made up of different combinations of just a few simpler objects. This situation would be analogous to the different atomic nuclei all being made up of different combinations of neutrons and protons. One such set of simpler objects which have been postulated are the quarks. To find out if quarks really exist one must knock them out of the elementary particles by high-energy collisions and then identify them in their free state. Accelerators have not been able to do this. The quark continues to elude the physicist.

Yet the mystery of the internal structure of the elementary particles was deepened by the results from the electron accelerators; results that are most easily explained by postulating some simpler objects inside nucleons (protons and neutrons). These results were obtained by colliding electrons or muons with nucleons on the accelerators at Stanford, Cornell and Cambridge in the United States; Daresbury, England; Hamburg, Germany; and Yerevan in the Soviet Union. The electron and the muon are

Plate 7.1 *The 3-km (2-mile) long electron accelerator at the Stanford Linear Accelerator Center, in California. It was in experiments in the late 1960s, in the "end stations" in the lower portion of the picture, that the first evidence for "grains" within protons and neutrons was gathered. The high-energy electrons were probes fine enough to begin to reveal the structure of the nucleons. (Credit: Stanford University)*

particularly interesting billiard balls to use, because as far as we know they are of infinitesimal size and have no internal structure. Furthermore, while most elementary particles interact with each other through both the nuclear (sometimes called strong interaction) force and the electromagnetic force, the electron or muon do not directly interact through the nuclear force; their major interaction is through the electromagnetic force. Now the electromagnetic force is relatively well understood, but the nuclear force poorly understood. Actually, our lack of understanding of the nuclear or strong interaction force lies at the root of our ignorance of elementary particle theory. Therefore in the collision of an electron or a muon with a proton we understand completely the role played by the electron or muon. In fact they come out of the collision unaltered in almost all cases. It is only the proton which is stirred up and which can be broken up into many elementary particles.

Thus we would expect the electron or the muon not to interact strongly with the proton and to slip past or through the proton as a needle slips through cork. Yet the experiments of the past two years show that energetic electrons or muons produce elementary particles quite readily when they collide with nucleons. It is as though there are hard kernels or pits in the nucleons which readily interact with the electron or muon. Therefore the collisions with these kernels – called partons – stirs up and excites the nucleon, giving it excess energy. In getting rid of this energy, the nucleon breaks up into several particles. Like the quarks, the partons have only been postulated.

Additional evidence for the parton hypothesis comes from the colliding-beam rings. In these machines, in which an electron collides with a positron, one would not expect to produce many pi-mesons. This is because a positron is just a positive electron and the major interaction of a negative electron with a positive electron will take place through the electromagnetic force. This electromagnetic interaction only leads to the production of pi-mesons by an indirect process. First, the electromagnetic interaction gives rise to one or more photons. Second, the photon or photons change into pi-mesons. But we expect this second step to occur with difficulty because, very roughly speaking, the photon is so different from the pi-mesons. However, if partons exist, they can act as intermediary particles greatly accelerating the conversion from photons to pi-mesons. Now preliminary results on the production of pi-mesons in electron–positron colliding-beam rings indicate that this production is much larger than initially expected. The partons provide one explanation, but there may be other explanations. The results are very new and the theorists have been taken by surprise.

17 September 1970

Monitor

To many physicists, the involved and expensive research into elementary particles has seemed increasingly to be leading to a dead end. Although the secrets of two of the four fundamental forces of nature, the "weak" and "strong" interactions respectively, are manifestly bound up in the behaviour of matter at high energy, progress towards unlocking them has appeared to many people to have become increasingly sticky. The 300 or so "particles" identified by the physicists at CERN, Berkeley, Serpukov, Batavia, and elsewhere may be nicely arranged into patterns of relationship, it is true, but the underlying significance of these patterns has remained elusive. The quark theory has, despite its workability, waned in popularity as successive experiments of many kinds have totally failed to find the faintest trace of the quark.

Now, however, the tide may be on the turn. For us ordinary mortals unversed in the mystique of particle theory it is hard to keep abreast of the developments as they happen. Ideally we need skilled interpreters. Fortunately one of the most skilled of all, Professor Richard Feynman of Caltech, has sounded an optimistic rallying cry for particle researchers, in a lecture given in Copenhagen. An issue of *Science* (vol. 183, p. 601) devotes nine pages to reporting his talk in full. It makes excellent reading. And, more importantly, Feynman reinstates the quark theory in all its glory. There seems every likelihood that physics is about to make another of its spectacular breakthroughs.

In the best tradition of physics, the thing that has caused the change of heart is experimental. For it now seems that, although they cannot be separated, apparently, it is nevertheless possible to see individual parts in protons which look remarkably like their constituent quarks.

The quarks consist of one having charge $+\frac{2}{3}$, designated "u", and two with charge $-\frac{1}{3}$, designated "d" and "s", respectively. The d and u quarks have equal mass and strangeness 0; the s

quark, which is responsible for contributing extra mass within a family of hadrons, has strangeness -1. From triplets of these three and their anti-quarks it is possible to build up all the known families of particles. Each meson appears as a quark—anti-quark pair.

So well does the quark model of hadrons work that so far no particles have appeared which cannot be explained by it. As Feynman says: "All this cannot be coincidence, yet one of the most obvious expectations is that these quarks should come apart in hard collisions between protons. Where are they? They have not been seen. They should be easy to see, carrying unusual charges like ⅔ as they do. . . ."

That the proton (the most familiar and accessible type of hadron) really does consist of three quarks (*u, u* and *d*) dancing around each other is borne out by measurements of its magnetic moment, a direct reflection of the alignments of the quark spins.

We have ample evidence that quarks can be exchanged in high-energy strong interactions. Indeed, such interactions can generate new quark—anti-quark pairs – mesons – out of energy. All that is necessary is that the net number of each type of quark remains the same before and after the reaction.

However, perhaps we can see these elusive creatures, if not alone, at least in company with one another. In short, does the proton consist of parts corresponding to quarks; and, if so, can we spot them in situ?

In fact, the experiment has been done on Stanford University's SLAC machine. The so-called deep scattering of very high-energy electrons by protons does indeed reveal such a quark-like structure. Feynman likens the scattering of point-like electrons from the proton's interior "to studying a swarm of bees by radar". In that case doppler shifting of the reflected radar frequency would show whether the bees were all moving in concerted fashion, or individually. The momenta of the scattered electrons are likewise a consequence of the momenta of the *charged* parts that scattered them. The results show that the electrons are apparently scattered by point-like parts – or "partons" – within the proton, and that this happens for various electron energies and scattering angles. The angular distribution of the scattered electrons also indicates that the proton parts have spin-½ in accordance with quarks.

This experiment does not tell us what the charges on the proton parts are, merely that there are distinct charged parts of the right spin, and that they are point-like. Another experiment is necessary to tell us the answer, this time bombarding protons with neutrinos which will interact with specific quarks and tell us something about their distribution. The trick is to compare scattering of the charged electrons from the partons, with that of the neutral neutrinos.

In the electron scattering experiment, the chance of an electron being scattered by a given part depends on the square of the charge of that part. Thus u quarks should contribute to the scattering with a weight of $4/9$, d quarks to the tune of $1/9$. Having made this assumption the data from SLAC can be compared with data from preliminary runs with the neutrino beam and the big bubble chamber Gargamelle at CERN. The two sets of data agree well, indicating that the partons do indeed carry the charges expected for the u and d quarks.

So we are left with imposing evidence for quarks without having seen them. Feynman is evidently strongly convinced of their reality. Have we perhaps reached the final barrier? Is matter at this level conceivably indivisible as the ancient Atomists believed? Such a view is anathema to today's physicists. So much so that Feynman is reduced to discussing an unlikely force law between quarks. Do they, he asks, attract one another with a force which remains constant whatever the distance apart of the quarks? If so they would require an infinite amount of energy to pull them apart – but perhaps that is not so very different from the Atomists' view.

21 February 1974

Monitor

Fresh evidence on the structure of protons and the force that holds them together has come from experiments studying the world's highest energy man-made particle collisions. The new results provide clear indications of collisions between the component parts of protons and are good confirmation of quantum chromodynamics (QCD), the theory of the strong nuclear force that not only holds together the constituents, but also binds protons with neutrons.

The experiments are analogous to the one in which Ernest Rutherford in 1911 first revealed the structure of the atom, with its hard central nucleus, much smaller than the atom itself and containing most of the mass. Rutherford and his colleagues at Manchester University fired alpha particles, which are in fact helium nuclei comprising two protons and two neutrons, at gold foil and recorded the directions in which they were scattered. Many alphas passed more or less straight through the foil but a few were scattered through wide angles. The scattering pattern Rutherford's team observed could be explained only if the alpha particles occasionally struck small hard nuclei.

Experiments over the past decade or so have provided similar evidence in the scattering of electrons, in particular, from protons, and this indicates that the proton contains smaller hard objects, "partons", which can scatter projectiles through wide angles. Physicists now generally believe that the partons are the elementary building blocks called quarks, and the associated gluons, transmitters of the strong nuclear force which flit between quarks so binding them together.

Colliding protons with protons allows physicists, in principle, to collide quarks and gluons with each other. Unfortunately, problems arise because quarks and gluons do not emerge from such collisions on their own but materialise instead as familiar particles, including protons, neutrons and pions. However, researchers have had some success in identifying "jets" in the debris of the collisions. These are sprays of particles which are believed to have a common origin in a quark or a gluon.

Jets are clearly visible, for example, in the products of the annihilation of electrons with positrons (anti-electrons). In this process the energy of the annihilation is believed to form a quark and anti-quark, which then materialises as two jets of particles. But the collisions between protons involve much more complex objects, and it is difficult to idèntify jets clearly in the resultant debris. However, earlier experiments at CERN's Intersecting Storage Rings (ISR), where two beams of protons collide head-on with a total energy of 63 GeV, had some success in identifying jets produced at wide angles, transverse to the beam direction, and these results suggested that scattering between hard structureless or "point-like" objects was occurring. But these experiments picked up only those collisions, or "events", in which one particle carried a large amount of momentum transverse to the direction of the colliding beams. The physicists then looked for evidence of jets in the opposite direction.

This method of selecting events with single particles carrying a lot of energy was open to criticism, so new apparatus has been designed that can detect the total energy of all the particles thrown out sideways. In this new experiment, physicists from CERN,

Scandinavia, Israel and the US have looked at those events in which a lot of energy, in some cases as much as 25 per cent of the total, goes sideways. They find that not only are there more events of this sort than they would expect were there no hard scattering but also that these events show jet-like streams of closely-associated particles, escaping sideways from the interaction region. The team has also been able to calculate the variation in the probability for the production of a jet with increasing transverse energy, and the results agree well with the predictions of QCD. The researchers are now installing more equipment to surround the collision region and they hope to acquire enough data to begin to distinguish between variants of QCD.

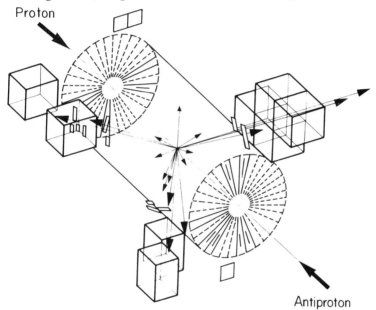

Proton

Antiproton

Figure 7.1 *A computer graphic display of a proton–antiproton collision in the UA1 experiment at CERN, shows three "jets" of particles leaving the collision zone and tracking out to various detectors. The lengths of the arrows indicate the momenta of the particles. The square blocks indicate that the jets contain strongly-interacting hadrons, which have triggered the hadron detectors. The formation of three jets, produced at wide angles with respect to the initial proton and antiproton, is indicative of a collision between hard constituents – quarks – within the incident particles.*

To see jets emanating from proton collisions really clearly, however, requires higher energies than the ISR can sample, and this is just what two experiments on CERN's proton–anti-proton collider have been doing. This facility enables physicists to study head-on collisions between protons and anti-protons with a total energy of 540 GeV, nearly 10 times that available on the ISR. Here the researchers have found the clearest examples yet of jets produced in proton collisions, in particular in those events with the most transverse energy (Figure 7.1). And not only are the jets clear to see, but they still occur at the rate that QCD predicts, even in this new region of high energies.

14 October 1982

8

The man who took quarks seriously

HARRY LIPKIN

Andrei Sakharov is confined to "internal exile" in the Soviet Union; his interest in quarks, which began with the first suggestions of their existence, continues despite his isolation.

In 1964, American physicists Murray Gell-Mann and George Zweig made the incredible suggestion that the supposedly elementary particles found in atomic nuclei, and in experiments at accelerators and with cosmic rays, were in fact composed of smaller constituents, called quarks. In those days it seemed absurd that the observed particles – specifically those belonging to two groups known as baryons and mesons – should be made of crazy new unobserved building blocks, carrying only fractions of the electric charge of the electron.

Physicists believed nucleons – the constituents of the atomic nucleus, this is, protons and neutrons – to be elementary particles, like electrons, because high-energy collisions did not break them up as they did molecules, atoms and nuclei, to reveal their constituents. Collisions left nucleons intact, producing a shower of particles called mesons, just as collisions left electrons intact and produced a shower of photons. According to the theory the Japanese physicist Hideki Yukawa developed in 1935, the mesons were interpreted in analogy to photons in electromagnetism, as the particles "exchanged" between nucleons – in other words, as the source of the strong nuclear force that holds the nucleus together.

At first, physicists generally ridiculed the quark model, but Soviet scientists Andrei Sakharov and Ya. B. Zeldovich took the model very seriously, and published a paper in the *Journal of Nuclear Physics (USSR)* (vol. 4, p. 395) in 1966. They, and independently P. Gederman, H. R. Rubinstein and I. Talmi in

Israel, found ways to test this model of unobserved quarks held together by unknown forces.

According to the model, quarks came in three types or "flavours" called "up", "down" and "strange", and had an intrinsic spin, like an electron has. All particles other than the electron and its relations contained the same quarks arranged in different ways. To study how the mass of the quark and the interquark forces depended on flavour, the two groups of physicists compared the masses of pairs of particles which differed only in the flavour of one of the constituent quarks; they studied the dependence of the force on spin by comparing the masses of pairs of particles which differed only by rearranging the orientations of the quarks' spins. Information from one such pair was used to predict differences in mass between other pairs. The predictions from such calculations agreed surprisingly well with experimental masses, thus providing evidence supporting the quark model.

Since 1966 many of the puzzles surrounding the quark model have been solved, particles containing very heavy quarks of two new flavours called "charm" and "bottom" have been discovered, and physicists now generally accept the quark model. The modern analogue of electromagnetism used to explain nuclear forces is no longer Yukawa's meson theory, but quantum chromodynamics (QCD) (see Part Four). The "chromoelectric" and "chromomagnetic" forces of the QCD theory hold the quarks together in a proton so strongly that they do not materialise in high-energy collisions. Instead new quarks are created and attracted to other quarks by very strong forces, so old and new quarks quickly find partners to make more mesons and baryons. Thus no isolated quarks are ever observed.

In 1975 Sakharov extended the application of the mass formula he had devised in 1966 to the newly-discovered particles that contain charmed quarks. In 1980, in exile in Gorki in central Russia, he improved the model using new ideas from QCD. I similarly updated the Israeli work from 1966 and obtained the same new results, in excellent agreement with experiment. The dependence on flavour and spin of the previously unknown interquark forces is now predicted by QCD. Chromo*electric* forces are independent of flavour and spin orientation, just as ordinary electronic forces are the same for electrons and protons. Chromo-*magnetic* forces are inversely proportional to the masses of quarks and depend in a well-defined way on the spin orientations – just as the ordinary magnetic force is 1000 times weaker on a proton than

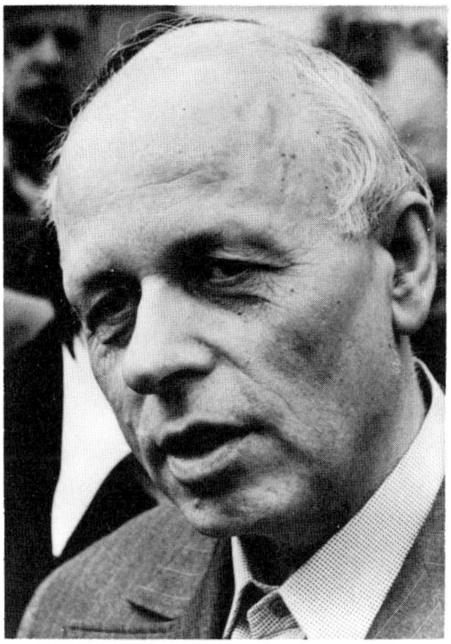

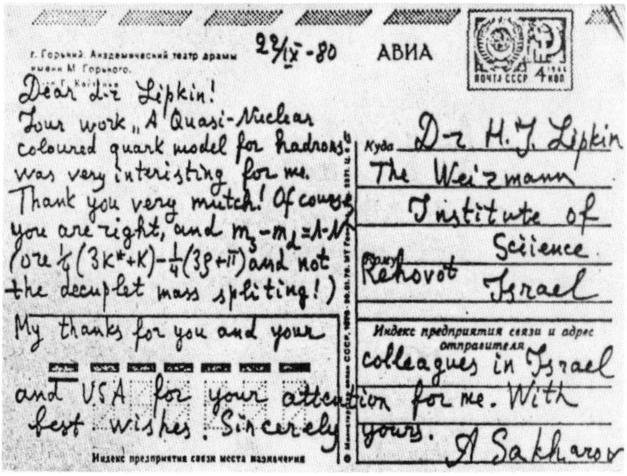

г. Горький. Академический театр драмы имени М. Горького.

22/IX-80

АВИА

ПОЧТА СССР 4 КОП

Dear D-r Lipkin!
Your work „A Quasi-Nuclear coloured quark model for hadrons" was very interisting for me. Thank you very mutch! Of course you are right, and $m_3 - m_2 = \Lambda - N$ (or $\frac{1}{4}(3K^*+K) - \frac{1}{4}(3\rho+\pi)$ and not the decuplet mass spliting!)
My thanks for you and your colleagues in Israel and USA for your attention for me. With best wishes. Sincerely yours. A Sakharov

Куда D-r H. J. Lipkin
The Weizmann
Institute of
Science
Rehovot Israel

Индекс предприятия связи и адрес отправителя

Индекс предприятия связи места назначения

Plate 8.1 *This postcard sent to Harry Lipkin in 1980 reveals that Andrei Sakharov, though in exile in the USSR, continued to work on theoretical particle physics, and in particular on the quark model which has interested him since its very early days. (Credit: Camera Press, Sven Simon)* on the much lighter electron, while also depending on spin orientation.

One example of a successful relation between the masses of mesons and baryons is the formula in the postcard that Sakharov sent to me from Gorki (see Plate 8.1). The formula was actually in his 1966 paper, obtained by an inspired guess before the theory QCD had been developed, but it had been unnoticed. The particle called the lambda is heavier than the proton: both contain three quarks but the lambda (Λ) contains one heavier "strange" quark. Similarly the K- and K*-mesons, made of one light quark and one strange quark but with different spin orientations, are heavier than the π and ϱ-mesons which contain two light quarks. Through QCD we know how to correct for chromomagnetic effects, and can predict that a suitable average of the K and K* masses is heavier than the analogous average of π and ϱ masses by exactly the difference between the lambda and proton. The striking agreement with experiment of this prediction confirms the hypothesis that mesons and protons are made of the same quarks.

Even more remarkable than these calculations of masses was Sakharov's use of the quark picture, together with a concept called "baryon nonconservation", to explain the excess of matter over anti-matter in the Universe. The world we know consists of protons, neutrons and electrons with no evidence for appreciable numbers of anti-protons, anti-neutrons and positrons. But in models of the origin of the Universe, equal numbers of particles and anti-particles are created. The accepted explanation in 1966 was that there must be anti-galaxies made of anti-matter somewhere in the Universe, even though no trace of such anti-matter is found in astrophysical observations. The only alternative explanation, that the anti-matter created in the big bang somehow disappeared in particle reactions and decay processes, was rejected. It violated the law of "baryon number conservation" which requires that if a baryon (such as a nucleon) is created or destroyed then an anti-baryon must also be created or destroyed – a law based on observations of particle reactions.

Baryon conservation was generally accepted to ensure the stability of the proton against decay into electrons and positrons and was supported by overwhelming experimental evidence. But even if baryon numbers were not conserved, baryons and anti-baryons have the same mass and must be present in equal numbers in any system in thermal equilibrium. Furthermore, conservation of "CP" – the combined effect of charge conjugation (changing particle to anti-particle) and parity (space reflection) – which is also observed in experiments prevents the creation of an excess of baryons over anti-baryons.

However in 1964, American physicists James Cronin and Val Fitch, and their colleagues, discovered that CP is not always conserved. Sakharov used this result together with violation of baryon number and the effects of a departure from thermal equilibrium, to explain the excess of baryons in the Universe (*ZhETF Pis'ma*, vol. 5, p. 32, translation *JETP Letters*, vol. 5, p. 24). He devised a model with a new, very weak interaction, which could turn the quark in the proton into electrons and neutrinos, with the emission of a new kind of very heavy particle called a "leptoquark gauge boson". Thus protons decay, but very, very slowly.

But at the time no-one took Sakharov's model with its novel propositions seriously: it assumed a quark structure for the proton; it postulated the existence of new kinds of particles – leptoquark gauge bosons – for which there was no evidence or need; it violated the conservation of baryon numbers and required the proton to decay; and it needed a departure from thermal equilibrium in the early Universe. Today physicists accept all these outlandish assumptions. Quarks are the basic constituents of the proton. The new grand unified theories of weak, electromagnetic and strong interactions introduce leptoquark gauge bosons and proton decay. The expanding Universe is not in thermal equilibrium, and goes through stages where the processes Sakharov suggested can occur.

30 April 1981

PART THREE

New quarks, new particles

November 1974 witnessed a revolution in particle physics. Two entirely independent teams of experimenters announced simultaneously the discovery of a new particle, the J/psi (the two teams could not agree on the name, so a compromise was made!). In some ways this event echoed the discovery of the strange particles in the late 1940s, which had heralded new concepts in theoretical particle physics. The J/psi did not fit into the patterns built from the three quarks Gell-Mann and Zweig had proposed. Robert Walgate charts these exciting times in the main part of Chapter 9, and presents the various explanations for the new particle that were aired in the weeks following the discovery. He also records the recognition given to the teams that found the crucial evidence, when in 1976 the leaders Burton Richter and Samuel Ting received the Nobel prize.

Walgate dwells longest on one particular solution to the puzzle of the J/psi's identity: that it consists of a new type of quark, carrying a new property called "charm", bound with its anti-quark. The "anti-charm" of the anti-quark cancels the charm of the quark, just as the charge of the positive nucleus and the negative electrons cancel in the neutral atom. This explanation soon became the favoured one, and Chapter 10 records the effort experimenters expended in searching for particles built from a charmed quark along with *uncharmed* quarks – in other words, particles with net charm, or "naked charm" to use the physicists' evocative terminology. The success of these experiments provides dramatic confirmation of the existence of a fourth type of quark.

But what of Gell-Mann's ideas? Did the discovery of a new kind of quark mean that Gell-Mann and Zweig were wrong? Like many questions in physics, the answer to this is "yes" and "no". Gell-Mann and Zweig were right in visualising the baryons – protons, neutrons and other "heavy" particles – as containing three quarks of any type; the Ω, for example, contains three strange quarks. Likewise, the picture of mesons – pions, kaons and so on – as quark–anti-quark pairs still holds. But Gell-Mann and Zweig worked with only three types of quark, which we now know to be a subset of the possible quarks. Thus their SU(3) symmetry based on three units is a "smaller" symmetry contained within the true symmetry group of all the quarks. Part Six will give some clues as to what that symmetry might be.

Meanwhile, Chapter 11 shows that there was more to follow the surprises of the J/psi. In 1976, experiments found evidence for a fifth type of quark, in the form of the so-called upsilon particle (Y). And as with the charmed quark, the search was set for particles containing only one quark of the new variety, called "bottom" for technical reasons, along with other quarks. Such "bottom particles" slowly revealed themselves, particularly in the 1980s.

The new quarks were not the only discoveries of the 1970s. Martin Perl describes in Chapter 12 the work of his team at the Stanford Linear Accelerator Center, in uncovering a new lepton, the tau (τ). Leptons are particles that are not built from quarks. They include, the electron, the muon (discovered in cosmic radiation in 1937, but at first mistakenly identified as the pi-meson) and the neutrinos. The neutrinos are neutral, probably massless particles, and there seems to be one type associated with each type of charged lepton. Thus physicists have identified an "electron-neutrino" and a "muon-neutrino"; it seems that a "tau-neutrino" must also exist.

So by the late 1970s a picture of matter had arisen that was quite different from the view held 40 years previously. Matter seems to be built from quarks and leptons. There are at least five kinds of quark and five properly identified types of lepton. In Chapter 13 Frank Close discusses present ideas that suppose there to be six types of each basic building block – six quarks, six leptons. He describes how nature seems to have created heavy "Xerox copies" of the four basic units needed to form the world about us – two quarks (to build the proton and neutron) and two leptons (the electron and its neutrino). And he will have more to say in Part Four, which looks at the "nuclear glue" that holds the quarks together in the many particles the chapters in Part Three have revealed.

9

A revolution in particle physics

ROBERT WALGATE

A new particle sets the world of high-energy physics buzzing and opens the way to a clearer understanding of quarks and the strong force.

There is a heady climate among particle physicists these days. To add to the lightness of the air, last week a new particle entered the scene. It was discovered independently by Sam Ting and his team at Brookhaven National Laboratory near New York, and Burt Richter and his boys (who are calling the particle a Ψ) at the Stanford Linear Accelerator Center in California. Physicists are comparing the result to the epoch-making identification of the strangeness quantum number by Murray Gell-Mann and Kazuhito Nishijima back in 1953.

The new particle is a heavy neutral meson. It has a mass of 3.105 GeV, roughly three times the mass of a hydrogen atom. That's twice the mass of the heaviest established meson. But what makes it really different is that it has an extremely long lifetime that takes it right out of the class of ordinary "resonances". These decay in times of the order of 10^{-24} seconds; but the ψ lives at least 1000 and probably 10 000 times as long.

At the time of writing its most interesting unmeasured properties are the details of its decay and its spin. So far it is known that it decays mostly into pions or kaons and occasionally into electron of muon pairs. The spin is limited to 0, 1 or 2. When these properties are known more precisely it will help theorists to choose between the various hypotheses for the ψ. These are that it is a meson composed of two new quarks (the "charm" quarks); that it is a Higgs particle (as postulated by an Edinburgh physicist Peter Higgs and crucial to unified theories of the weak and electromagnetic interactions); that it is a heavy version of the photon;

Plate 9.1   *Burton Richter (left) and Sam Ting (right) shared the Nobel Prize for Physics in 1976 for their discovery of the J/ψ particle two years earlier. (Credit: SLAC and BNL)*

and that it is the intermediate vector boson of the weak neutral current (see Part Five).

First signs at Stanford of the particle came in June, when Richter's electron–positron colliding beam machine SPEAR indicated a reaction rate at energies near the mass of the ψ hundreds of times higher than at neighbouring energies. Afraid that this was some problem in the equipment Richter's group spent months checking the result. According to Richter, last week the group decided the ψ meson was real.

Sam Ting at the Brookhaven machine (an old 33 GeV proton synchroton) had seen the same meson in collisions of protons with beryllium nuclei in which electron pairs were produced. Ting has been working painstakingly on similar reactions for 10 years. Some months ago he saw a bump in his graphs where the ψ now sits, and he, like Richter, checked and rechecked. By last week Ting had decided the particle was there, and excitedly called on Richter. At that point they decided to inform the world.

MONITOR
21 November 1974

Monitor

SPEAR, the electron–positron colliding beam machine at Stanford, has driven home yet again with another new particle. This one is like the ψ (psi) discovered last week but has a higher mass (3.7 GeV as opposed to 3.1). Its lifetime is again long, which this time brings another problem: if the new state is just an excitation of the lower one (the most economical hypothesis) why doesn't it decay rapidly to the lower state?

The heavier particle was found last week after only 6 hours search. The researchers started with the machine tuned to 3.6 GeV and slowly wound up the energy. According to Burton Richter, the director of SPEAR, the whole region from 3.1 to 3.6 GeV has – for technical reasons – yet to be covered and that they may find even more particles in that intermediate band. Richter said of the new particle "normally I would have liked to sit back and work things out before I announced it"; but, he said, the news leaked out to local theoreticians and before long it was common property.

SPEAR is now developing its investigations along three tracks simultaneously: looking at the details of the decay of the two particles and the search for further partners for them. For each particle the machine has to be separately tuned and the work is becoming very time consuming. "What we really need right now" said Richter "is another electron–positron storage ring".

In fact there is one. It's an Italian machine called ADONE at Frascati near Rome, and it has just enough energy to reach the lower mass ψ at 3.1 GeV. Confined to this particle, the physicists at that machine have looked carefully at its decay. They have found it to be very complicated. The mean number of charged particles in the decay is 3.4 and they have seen a maximum of 8; the mean number of gamma rays in each decay is 1.6 and they have seen a maximum of 7. Such a complex decay is reminiscent of the decays of the higher mass resonance particles; but the critical contrast is that the decay rate is so slow. Both ADONE and SPEAR have been trying to measure the spin of the lower state. Although no firm measurement has been announced the early indications are that the spin is not zero. If this were confirmed it would rule out the hypothesis that the ψ is a Higgs particle.

28 November 1974

Psi story

Two new particles have been discovered by the high-energy physicists. They are the first stable particles to be discovered since the Ω^- of 1964. But whereas the Ω^- was a triumph of theoretical prediction, the new particles (we call them provisionally the ψs) are a success for the experimenters and a trouble to the theorists. Over the past few years particle physicists have been coming to understand their subject for the first time since the wretched "V particles" (now well labelled and pigeon-holed and called strange particles) rocked the boat in 1948.

Out of this understanding new particles have been predicted going by such lovely names as intermediate vector bosons, Higgs particles, charmed and coloured particles and heavy light. So the first flush of excitement – amongst the theorists – was the thought that the ψ particle announced on 11 November (now there are the $\psi(3105)$ with a mass of 3105 MeV and the $\psi(3695)$ with 3695 MeV) fell into one of these slots so carefully prepared. The second flush – this time of embarrassment – came when it was seen that the ψs (by then there were two) didn't properly fit into any of them. The initiative has returned to the experimenters to pin the ψs right down and provide the theorists with the last detail of their anatomy.

Yet theory progresses by successive refinements in response to new data, and it may be that the adjustment the ψs require to one of those theoretical slots will just nudge theory towards a complete picture of the elementary particles. This is not so far fetched a possibility as it may seem; one of the slots would have the ψ composed of two new quarks, charm quarks, which have been postulated by theorists. On the other hand, of course, the ψ could turn out to be such a bad fit to expectations that all previous ideas have to be cast aside.

The background to the ψ story is success, setback and (over the past 3 years) greater success in the understanding of elementary particles. In the early 1960s the subject was growing fast, driven by bubble chamber experimenters – the people who take pictures of reactions by photographing trails of bubbles in liquid hydrogen – who were discovering more and more "resonances". These were principally excited (higher mass) states of the basic particles of the nucleus, the proton and neutron and the pion (which holds the nucleus together) and their strange particle counterparts: sigmas, lambdas and the kaon. The resonances decayed very rapidly (in 10^{-23} seconds or less). In 1964 they were neatly grouped together by the hypothesis that they were all composed of even more fundamental entities – the quarks – and the Ω^- was observed in a copybook confirmation of the "unitary symmetry" underlying the quark theory.

In the past 3 years the quark picture has been immensely strengthened by measurements of electron scattering from protons which showed clearly that the proton behaved exactly as if it contained three point-like objects with just the right (third-integral) charges predicted for the quarks. The free quark has not been discovered but that is a problem that is being increasingly ignored as unimportant.

Broadly, the results of the synthesis have been that there are three basic interactions between particles and three classes of particle: the photon, the particle of light, in a class on its own with only the electromagnetic interaction; the leptons, just four particles at present (and their anti-particles), with electromagnetic and weak interactions (the weak interaction causes beta decay in nuclei and controls the heat output of the sun); and the hadrons, composed of quarks, a large class of particles with the electromagnetic, weak and strong interactions (the strong interactions bind the nucleus together). The hadrons are further divided into mesons composed of quark and anti-quark (like the pion) and baryons composed of three quarks (like the proton). All the particles can be divided between two classes, "fermions" and "bosons". Fermions have half integral spin and bosons integral spin. The principal mode of interaction between particles is of fermions firing bosons at each other, feeling the recoil like machine gunners; the strength of the interaction determines the rate of firing. The photon is the "bullet" for the electromagnetic interaction, and the pion can be regarded as a "bullet" in the strong interaction.

The only interaction where there was no natural boson "bullet" was the weak interaction. A theory due to Steven Weinberg and Abdus Salam which unifies the weak and electromagnetic interactions predicts there be a boson for the weak interaction: the so-called intermediate vector boson (IVB; vector means spin 1). It would come in three varieties, positive, negative and neutral. The $\psi(3105)$ is definitely a boson and it could be the neutral IVB. However, its mass is too low for the Weinberg—Salam theory, and that theory would have to be radically modified (it might not remain a unified theory) if the $\psi(3105)$ were indeed the neutral IVB. Also the existence of the $\psi(3695)$ and its subsequent decay into the $\psi(3105)$ and two pions have no place in the theory.

Most likely the particle physicists will want to abandon the idea that the ψ is an IVB and retain the Weinberg–Salam theory. But that theory and others of the same ilk (the "gauge theories") predict yet other particles – the Higgs particles – and hint at others – the charmed particles. Higgs particles have a rather ghostly status in gauge theory; they are necessary to make the theory calculable but may be no more than theoretical devices and be unobservable. They would have spin 0 and interact rather weakly with electrons, and as in the IVB hypothesis for the ψ it is difficult to understand in the Higgs hypothesis how the decay of the $\psi(3695)$ into two pions would work.

Slightly more likely is the picture that the ψ is connected with charm. Unified field theories of the weak and electromagnetic interactions work with two "doublets" of particles – the electron and electron neutrino, and the muon and muon neutrino – and would mesh more neatly with the strong interactions if there were two doublets of quarks. The required fourth quark has been called the "charm quark".

In the standard quark model, the hadrons (the vast majority of the particles – those feeling the strong interaction) are composed from a lucky dip of three basic pointlike objects, the u, d and s quarks. The baryons (particles like the proton and the neutron) pick three quarks from the dip (for example the proton is composed of two u quarks and one d quark), while the mesons (particles like the pion) pick two – one quark and one anti-quark.

The three quarks are distinguished by just two internal qualities "hypercharge" (related to "strangeness") and the "third component of isotopic spin" (related to electric charge). For short these are called Y and I_3. It is possible to keep a numerical account of these qualities – to turn them into quantities – because their values are conserved (in other words obey simple arithmetic) in the strong interactions. It is thus possible to plot the quark content of any hadron on a two-dimensional graph of Y against I_3. This is done in Figure 9.1 for the spin-1 mesons. There are 3 quarks and 3 anti-quarks which allows 9 mesons (a nonet); they are the 4 K*s, the 3 ϱs, the ω and the φ.

Now the new $\psi(3105)$ probably has spin 1 but it doesn't fit into this pretty nonet. So one popular explanation of the new particle is to recall weak interaction theory which could describe weak inter-

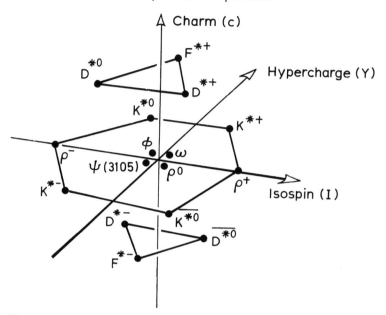

Figure 9.1   *The quark model based on three types of quark predicts nine mesons with spin-1. These fit on a hexagonal plot of isospin versus hyperchange, with three mesons, the rho (ρ), omega (ω),. and phi (φ), at the centre. With four basic quarks we need a three-dimensional plot, to include the new property of the fourth quark – "charm". Then we find we can make 16 spin-1 mesons: two sets of three with charm, and 10 without charm. These latter include the original nine mesons and an additional "charmless" particle built from a charmed quark bound with its anti-quark – in other words, the ψ (3105), now better known as the J/ψ.*

actions of the hadrons more neatly if there existed a fourth quark in the lucky dip. This quark has been called the "charm" quark, c. In this model the $\psi(3105)$ would be composed of a charm quark and an anti-charm quark, $\bar{c}$. The strong interactions would conserve charm so it too would be quantifiable, and the quark content of hadrons could be plotted in a three-dimensional graph of charm against Y and I_3. There would be 4 quarks and 4 anti-quarks, giving 16 mesons. Nine of them would be the familiar ones with no charm quark present; there would be three with one charm quark, three with one anti-charm quark, and one – the $\psi(3105)$ if this model is right – with one charm and one anti-charm quark.

(The $\psi(3695)$ would have to be an excited state of the $\psi(3105)$ and the decay of the $\psi(3105)$ would have to be inhibited for some reason.)

The 16 spin 1 mesons are shown in Figure 9.1, with the charmed triplets above and below the orthodox, zero-charm plane. There would also be 16 lower mass, spin 0 mesons fitting into an identical picture with similar quark content but with the quarks in an orbital state with one unit less of angular momentum. In this 16 the zero-charm nonet would comprise the well established 4 Ks, 3 πs and two particles called η and η'. There would be another charmed neutral η_c but it would not be seen in electron reactions and would be swamped in others.

The test of this four-quark model will be to find the predicted charmed triplets, the D* and F* mesons of spin 1 and the D and F of spin 0. Knowing the mass of the $\psi(3105)$ it is possible to calculate the masses of the D* and F* as about 2.2 GeV and the masses of the D and F as about 1.9 GeV. The D* and F* would decay rapidly (lifetimes about 10^{-23} seconds) into the D and F. These would in turn decay (but more slowly) into other hadrons; for example the D^0 would decay into a kaon and a pion with a lifetime of about 2×10^{-13} seconds.

There is one problem with the ideal that the $\psi(3105)$ is a charm–anti-charm meson – its lifetime is too long (by a factor of 100 or so). However, exact calculation of the lifetime is not possible, and this – so say the optimists – may prove to be a creative rather than a destructive problem.

Another charming niche for the ψ is the D^{*0}, a charmed meson with only one charm quark. In this case (as charm is conserved in the strong interactions) the Brookhaven experiment must produce two ψs simultaneously: one the anti-particle to the other. The experiment has only just enough energy to do this for the $\psi(3105)$ and not enough for the $\psi(3695)$, so the hypothesis would be an explanation for Brookhaven's not seeing the heavier particle. However, the lifetime would then be wrong by a factor of 1000 or so. A test of the idea is in progress at the moment; the proton beam energy is being turned down so low that the $\psi(3105)$ could be made if it came alone, but not if it came in pairs. If it disappears from the plots it follows that it is made in pairs and the D^{*0} hypothesis will reign supreme.

There are three other possibilities: that the ψs are coloured particles, or heavy photons, or that there is no slot for them whatever in existing theories. "Coloured" particles invoke the ideas of Han and Nambu that the normal ("white") hadrons are built from a kind of average over "red", "green" and "blue" triplets of quarks. The idea has many varieties; it is designed to deal with technical difficulties in the quark model. Some varieties predict two neutral bosons that couple to the photon. They would be produced in the experiments that saw the ψs, but it is difficult to see why they have not been produced before in interactions between photons and protons. However, the coloured quark model has many faces, and there may be some variety of it that fits the facts.

The heavy photon (heavy light), the last theoretical refuge for the ψs, was invented by Lee and Wick to resolve once and for all the infinities that are brushed (successfully) under the carpet of quantum electrodynamics (the exact and highly successful relativistic quantum theory of electromagnetism). If it existed it would be possible to calculate quantities like the mass difference between the proton and the neutron. Unfortunately, it would interact with the electron in the same way as the ordinary photon; and in the best interpretation the ψ has only half that strength of interaction. But a factor of two might be easily corrected.

The last possibility – that ψs are entirely new – is the most disturbing. That indeed could mean the unravelling of the whole subject.

19 December 1974
and Monitor

Box 9.1   Burton's bonanza

In Burton Richter's electron and positron storage ring at Stanford, SPEAR, intense beams of electrons circulate in one direction and positrons in the other. The beams are kept separate except at pre-arranged collision regions where experiments are done. The energy of the beams is fixed for one measurement but may be changed between totals of about 0 and 7 GeV (roughly 0 to 7 proton masses, using $E = mc^2$).

Diagrams of typical processes that have been measured over the past 18 months are shown in the figure. The electron and positron collide and create other particles, but only charged particles can be detected. (Thus the reactions with all neutral hadron production are unobserved.) All the previously known processes can be described by assuming the electron and positron collide and convert into a photon γ (a particle of light) by a process that can be precisely calculated using the most accurate theory ever invented, quantum electrodynamics (QED). Energy and momentum are conserved at the vertex.

Box 9.1 *continued*

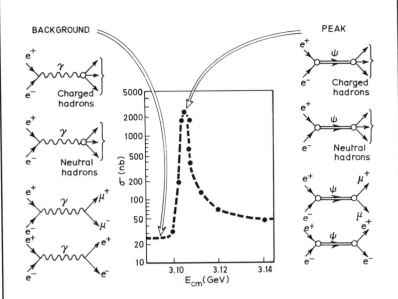

At Stanford, Burton Richter's team found a peak in electron–positron $(e^+ e^-)$ collisions at an energy of 3.105 GeV, where the counting rate rose 1000 times above the background.

Subsequently the photon converts by an unknown process (indicated in the diagrams by an open circle) into charged and neutral hadrons (mostly pions) or by a calculable, QED process into positive and negative muons (heavy electrons) or an electron and a positron. Hadronic interactions can be distinguished experimentally simply by there being three or more particles produced – a process which in QED is less likely than the two particle process by a factor of about 137.

At any energy, these processes take place quantum mechanically: they occur one at a time with certain probabilities but randomly. The probability of each reaction is measured as a counting rate for that reaction in the detectors. These photon-mediated processes (left hand column), which have made up most of the SPEAR program since the machine was built, constitute the background to the production processes for the ψs (right hand column).

continued

Box 9.1 *continued*

Conservation of energy indicates that a ψ will only be produced when the sum of the energies of the electron and positron is equal to the mass of the ψ. As can be seen in the central graph of electron–positron "cross section" (a measure of total counting rate) against energy, there is an enormous peak at 3.105 GeV as the beams tune in to the ψ – a peak a factor of 1000 above the background which must be plotted on a logarithmic scale to get the signal on the page.

Box 9.2   Samuel's success

Samuel Ting's experiment at the Brookhaven National Laboratory was in a sense the reverse of the experiment at SPEAR. A very intense proton beam was fired at protons (in a beryllium target) and measurements made of particles produced in the collision. Only one in a million of these particles were electrons, but with great ingenuity Ting's team was able to separate them, look for pairs (a simultaneous electron and positron) and for each pair calculate the mass of the object they came from.

After collecting a great deal of data, the team could plot how many "events" appeared in every interval of mass, say between 3.000 and 3.025 GeV (where there were 10 events). The section of the plot around the $\psi(3105)$ is shown here, with the large peak at the ψ mass of 3.105 GeV. The photons which compete with the ψ to make electron–positron pairs can come at any mass and they make up the background.

Although the workers at Brookhaven had already seen the peak in August of this year, it was so dramatic they felt obliged to check and recheck their results. One test they did was to run their magnets at 10 per cent lower current. If the ψ peak had shifted it would have been an artefact of their apparatus. As can be seen from the diagram, it didn't.

Box 9.2 *continued*

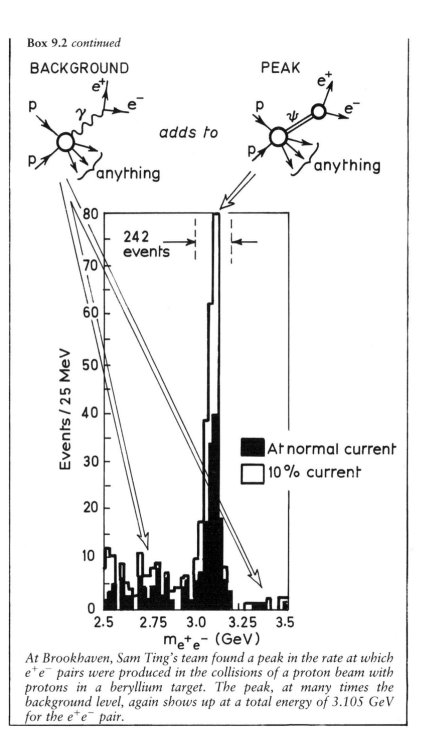

BACKGROUND

PEAK

adds to

At Brookhaven, Sam Ting's team found a peak in the rate at which e^+e^- pairs were produced in the collisions of a proton beam with protons in a beryllium target. The peak, at many times the background level, again shows up at a total energy of 3.105 GeV for the e^+e^- pair.

Nobel prize

On 11 November, 1974, Samuel Ting, professor of physics at the Massachusetts Institute of Technology, walked into the director's office at the Stanford Linear Accelerator Center, California, and said: "I'd like to talk a little physics." Ting had discovered a new particle with very unusual properties. He called it the J and had found it in his experiments at the Brookhaven National Laboratory, Long Island, over 3000 km away on the other coast of the United States. Wolfgang Panofsky, somewhat surprised, told Ting that one of the SLAC group leaders, Burton Richter, had just discovered a similar particle at Stanford. Richter's experiment had been completed in the two days before Ting's visit, Panofsky said, and the brief paper announcing the result was already written.

Ting and Richter compared notes and soon convinced each other that they had each discovered the same particle. Richter had called it the psi (ψ). He and his colleagues had gone through the list of elementary particles and marked off the letters of the Greek and Roman alphabet already used. They were left with psi and J, but they rejected J as it is the usual symbol for angular momentum. The particle now goes by the name of J/psi.

For the discovery of this particle Richter and Ting have been awarded the Nobel Prize in Physics for 1976 – one of the swiftest accolades on record from the cautious Nobel Foundation. The Foundation could respond quickly because the existence of the particle is in no doubt; and because the particle and its subsequently discovered relatives have given a tremendous boost to high energy physics.

What is the J/psi? From early on Ting himself took an iconoclastic view swimming strongly against the mainstream, while Richter remained open-minded. However, even Ting now believes that it consists of a new quark, the charmed quark, bound in orbit to its anti-quark. The existence of the J/psi has given support to a rather simple picture of the world. According to this view, matter is made by two kinds of particle: quarks and leptons. These particles are point-like (in other words have no discernible structure) and interact by means of four kinds of force. The forces are the strong interaction (which binds the nucleus together); the weak interaction (which occasionally prises the nucleus apart); the electromagnetic interaction (which binds electrons to the nuclei of atoms and creates light); and gravity.

The difference between leptons and quarks is that leptons are blind to the strong interaction – they do not react to it. As the strong interaction is the strongest of all forces it dominates the behaviour of the quarks; it is so strong that quarks cannot even unbind themselves from the particles – the "hadrons" – in which they are found. For example the familiar proton and neutron are hadrons (made of quarks); the electron, on the other hand, is one of the leptons.

Until the J/psi, only three quarks were known: the up, down and strange quarks, which differed in their masses and decay properties. But the lifetime of the J/psi found by Richter and Ting was 10 000 times too long for it to be composed of the three known quarks. If, on the other hand, it were composed of new quarks that could decay to the old ones only with difficulty its properties could be explained. The new quark was called the charmed quark (the name already existed) and physicists needed four quarks to describe the world. As they also needed four leptons (the electron, the electron neutrino, the muon and the muon neutrino) that indicated the existence of a simple kind of symmetry in nature between quarks and leptons.

It was a symmetry that had already been invented, on grounds of elegance, by theoretical physicists. The symmetry found its fullest embodiment in a unified theory, created independently by Abdus Salam and Steven Weinberg, of two of the forces that the leptons and quarks have in common – the electromagnetic and weak interactions.

The behaviour of the J/psi thus fits neatly into a slot provided for it by theorists. There were competing ideas but by today, less than two years later, a dozen or so related particles have been discovered (see Chapter 10). All show the properties expected of particles containing charmed quarks (as predicted by the quark–lepton symmetry idea) and so the charm interpretation of the J/psi is firm.

Of the two discoverers of the epoch-making particle, Samuel Ting is the younger. He was born in 1936 at Ann Arbor in Michigan, but spent most of his youth as a refugee in China. He returned to the US at the age of 20, and earned a PhD from the University of Michigan in 1962. He spent a year at CERN, two years as an assistant professor at Columbia University, and then in 1966 began his career proper at the Deutsches Elektronen-Synchrotron, DESY, near Hamburg. There he immediately began studying the production of leptons. At Brookhaven, eight years later, he was investigating lepton production in hadron collisions –

an extremely taxing job as hadron collisions produce mostly hadrons – when he found the J/psi by its decay into electron pairs.

Richter was born in Queens, New York, in 1931, took his PhD at the Massachusetts Institute of Technology in 1956, and headed straight for Stanford University (which at the time had a 1 GeV electron accelerator) determined to scatter electrons from electrons. Richter hoped to find a physical cause for the abstract "renormalisation procedures" that theorists used to remove the infinities from quantum electrodynamics (see Chapter 18). In 1957 Richter was already talking to Panofsky of a storage ring to perform such experiments. Ultimately Richter created SPEAR, a storage ring for electron—positron collisions. SPEAR found the J/psi and most of the subsequent charm particles; in the hands of Richter and his 40-strong team of colleagues it has been the most creative experiment in the history of particle physics.

21 October 1976

10
Particles with charm

The new generation of heavy particles containing the fourth "charmed" quark, proves a challenge to the experimenters.

The possible discoveries of a charmed baryon and a charmed meson (counterparts of the proton and pion) were announced in totally different styles by a group from CERN and a group from Brookhaven at the International Colloqium on Neutrinos held in Paris last week.

The charm that the experimenters may have found is, as one member of the CERN group put it, "a theorist's dream". Charm is the hypothesis of another kind of conserved charge for the strong interactions, whose existence would explain a number of theoretical problems. Not least of them would be the existence of the new psi mesons discovered in 1974: the psis would have zero overall charm but be composed of one charmed quark and one anti-charmed quark which are spinning around one another.

So, fired with enthusiasm, one group, Nick Samios's team at the Brookhaven National Laboratory (BNL) – represented in Paris by the group's co-leader Robert Palmer – threw caution to the winds and announced their event (for a tentative charmed baryon) with full pomp and ceremony. But the other group – the Gargamelle bubble-chamber team from CERN, headed by Paul Musset – played down their tentative discovery (the meson).

The charm candidates are two neutrino events, the one recorded in CERN's huge Gargamelle heavy-liquid bubble chamber, the other in the new 7-foot cryogenic chamber bubble at Brookhaven. Both events attract attention because they violate a sacrosanct rule of charmless theory – that when protons are struck by neutrinos the resulting excited state changes electric charge by the same

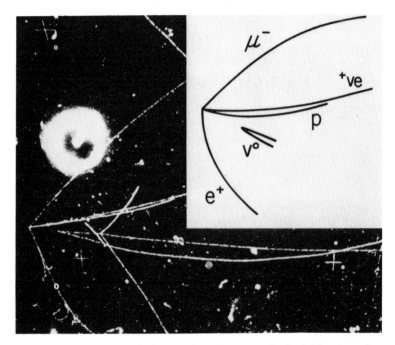

Figure 10.1 (a) *Tracks in the Gargamelle bubble chamber at CERN which indicate the decay of a charmed particle. The production of a positron (e^+) and a neutral strange particle, revealed by its decay (marked V^0 in the top picture), is the hallmark of the decay of a charmed particle. (Credit: CERN (top))*

amount as it changes strangeness (strangeness is another kind of charge for the strong interaction).

In the event from BNL, however, the proton (charge $+1$ and strangeness 0) changes to a Λ (charge 0 and strangeness -1), plus four pions (total charge $+2$ and strangeness 0). Thus the strangeness has gone down one, but the charge has gone up one (the neutrino changes to a negative muon, conserving charge overall).

In the event from CERN a similar violation takes place. Again the neutrino changes to a negative muon, so the charge on the proton system rises by one, but this time a K^0 meson is produced. Superficially it is impossible to tell what strangeness the K^0 carried (it can be $+$ or -1) but its production is associated with a positron. Cognoscenti of charm physics can then guess that its strangeness was -1 and that there may have been the decay of a charm meson.

CERN's event would then be the production of a singly charged charmed meson decaying into a positron, a neutrino and a K^0. Brookhaven's event would be a charmed baryon decaying into a Λ and some pions. Palmer, indeed, went so far as to hint at the mass of his charmed baryon. If all the pions constituted its decay product, the mass comes out to be 2424 MeV with an uncertainty of 12 MeV. Its charge would be $+2$, and this just happens to fit the mass of one of the charmed baryons predicted from the slime psi – viz the hypothetical Σ_c^{++} with a mass of 2423 MeV. It was Samios's group that found the Ω^- at Brookhaven in 1964 (see Chapter 3). But the Ω^- was a more clearly signed event, with a measurable Ω^- track in the chamber; charm particles decay too fast for that so their identification is more difficult. The discoveries at Brookhaven and CERN must be treated as tentative until there is a larger collection of such events. We shan't know for some time whether Brookhaven's trumpet blowing, or CERN's caution, was the wiser approach.

MONITOR
27 March 1975

Monitor

Elementary particle physicists are gearing themselves up for what may turn out to be the final onslaught on the bastion of "charm". An experiment using a neutrino beam and a heavy-liquid bubble chamber, although only in the early stages of analysis, has already yielded four very special events which could include charmed particles. Jack Fry, Professor of Physics at the University of Wisconsin and spokesman for the Wisconsin–Berkeley–Hawaii–CERN collaboration that performed the experiment at the 400 GeV Fermilab near Chicago, said last week that at the rate at which the events are turning up at least 20 and perhaps as many as 50 will be collected by the time the analysis is complete.

The discovery is particularly exciting as it confirms the findings of CERN's Gargamelle team which announced one event of the same type at a conference in Paris several months earlier. Fred Bullock of University College, London, who works with Gargamelle, says that the group has found a second event, and is drafting the paper that would announce it. Gargamelle has, and

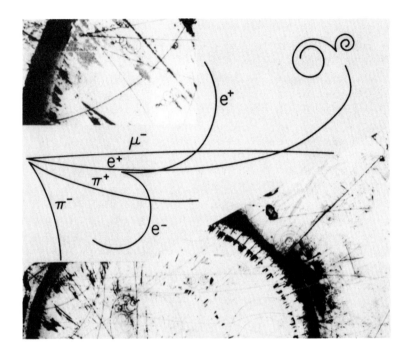

FIGURE 10.1 (b)   *Tracks in the 4.5 m chamber at Fermilab which indicate the decay of a charmed particle. (Credit: Fermilab)*

probably will have, fewer examples to show than Fermilab because its neutrino beam is restricted to a maximum energy of 30 GeV; Fermilab can reach up to 400 GeV or so. Production rates almost certainly increase with energy.

The existence of charmed particles would be the strongest indication that the physicists' most ambitious picture of the world of the elementary particles, that knits together the weak, the electromagnetic, and to some extent even the strong interaction, is essentially correct. Charm would also explain the psi particles discovered a year ago.

The events in question (Figure 10.1) appear to be

$$v + \text{nucleus} = \mu^- + K^0 + e^+ + \text{hadrons}$$

followed by the normal decay of the strange particle

$$K^0 = \pi^+ + \pi^-$$

Single positrons (e^+) emanating directly from the collision are uncommon for two reasons: first, the incoming muon-neutrinos cannot produce them (they can only make muons); secondly being leptons subject only to the weak and electromagnetic interactions, the positrons cannot be directly produced by the excited nuclei, either (the nuclei composed of hadrons, are dominated by the strong interaction). So when a single positron is produced it generally indicates that a second weak interaction has taken place in the nucleus – after the weak interaction with the incoming neutrino that first excited the nucleus. What happens is that one of the hadrons produced by the nucleus decays weakly into a positron, an electron-neutrino, and another hadron.

But the lifetimes of all the common particles (the strange particles) that decay in this way are so long that they would have left clear tracks in the bubble chamber before they departed this life. So here the charm interpretation makes its entry. Charmed hadrons are expected to be so massive (more than twice the mass of the proton) that their weak lifetimes would be a thousand times shorter than those of the strange particles and their tracks correspondingly immeasurably short.

Then comes the final confirmation: charmed hadrons are expected to produce a strange particle as their predominant hadronic decay product – and, lo and behold, with each of the four directly produced positrons found so far at Fermilab, the decay products of a strange particle – the K^0 – are also to be found. (The Gargamelle team makes the presence of such a particle a condition of the event's acceptance.)

This then is the signal that both CERN and Fermilab now seem to have seen. It will be only a matter of time before the few tens of examples that Jack Fry and his friends hope to collect will tell us for sure whether the signal is truly charmed.

18/25 December 1975

Monitor

A nakedly charmed particle has been revealed at SPEAR, the US storage ring designed to collide electrons with positrons. The news comes 18 months after the discovery of the psi particle, which first raised hopes that charm was more than a theorist's dream. If the most popular model is right, the new particle consists of two quarks one of which carries the special "charge" called charm. The psi particle in this model contains both a charmed quark and its anti-particle so that it exhibits no net charm. The new particle, on the other hand, would have a net charm of one. The particle found at SPEAR is neutral and has a mass of 1.87 GeV. It has been found by summing all the data for electron–positron collision energies between 3.9 and 4.5 GeV, where the interaction rate between electrons and positrons has been known – for nearly a year – to show lots of variations. This structure was always thought to have something to do with the production of charm.

The particle appears to be made (as it should) in association with another particle or particles (with the opposite sign of charm) whose masses can be determined from the spectrum of mass recoiling against the 1.87 GeV object. Two tentative states, at 2.02 and 2.15 GeV, have been identified in this way. The 1.87 GeV particle decays into a K-meson and a pi-meson, just as it should if it were a charmed meson. Its decay into a K-meson and three pi-mesons has also been detected. Its lifetime has not yet been determined, but its mass is just about as predicted.

20 May 1976

Monitor

An experiment headed by Professor Eric Burhop of University College, London, appears to have found the first tracks showing the production and decay of a "charmed" particle. The tracks were formed by an event induced by a neutrino in photographic emulsion exposed to a high-energy neutrino beam at the Fermi National Accelerator Laboratory, Chicago. Other experiments at the Stanford Linear Accelerator Center and at Fermilab have indirectly demonstrated the existence of charm – a new kind of "charge" in nature – but none until this one has recorded the track left by a charmed particle as it passes through matter. Charm, which first turned up in November 1974 (in the shape of the J/psi particle), has brought great harmony to our understanding of the nature of matter.

The problem has been the extremely short lifetime of charmed particles – predicted to be of the order of 10^{-13} seconds. Bubble chambers, which leave tracks of vapour bubbles in a superheated liquid, can resolve distances of only 100 to 200 micrometres. Spark chambers and other electric detectors have even worse resolution. Photographic emulsion, on the other hand, can resolve a single micrometre and thus has the capability of detecting a charmed track.

Burhop, after experience with a similar experiment at the CERN laboratory in Geneva in 1964, decided to expose a 20-litre stack of emulsion to a neutrino beam at Fermilab. The emulsion stack was combined with an array of spark chambers. Tracks recorded electronically in the chambers were to be followed back by computer into the emulsion stack to pinpoint the place (to within a cubic centimetre) where the initial neutrino interaction took place. This would reduce the time taken to find the interactions in the emulsion from 1000 to 10 or so man-years.

Burhop's team found a potential charm candidate in what was only the fifth interaction scanned in the emulsion. The composite micrograph of the event is shown in Figure 10.2. A neutrino, leaving no track, enters from the top, splatters a nucleus in the emulsion to fragments (heavy tracks) and creates some light tracks. One of these (the track AB) moves forward and within 182 micrometres (half way along the picture) decays into three charged

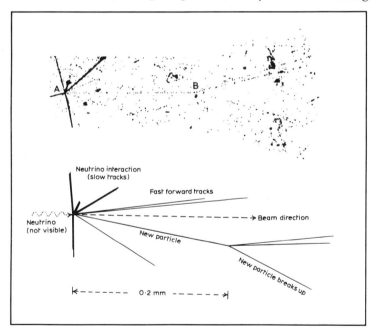

Figure 10.2 *Tracks in emulsion exposed to a beam of neutrinos at Fermilab show the formation and decay of a new particle, which leaves a track marked AB. All the evidence suggests that the track was made by a charmed particle produced in the initial interaction.*

tracks. Any explanation of this event other than the production of charm has a probability of only 1 in 1000 or thereabouts of being correct.

The potential charmed particle (track AB) has a lifetime of the order of a few times 10^{-13} seconds. The precise lifetime of the particle depends on the exact velocity. This is not known but can be estimated.

Burhop says he would not have published the event but for one fact – it has a neutral decay associated with it. Charmed particles are expected to decay predominantly into "strange" particles, which, in turn, decay into ordinary particles. About half the common strange particles (K-zero and lambda-zero particles) are neutral and decay in a metre or so into "V" tracks consisting of oppositely charged particles. A month after the discovery of the potential charmed event, Burhop's team discovered a "V" track associated with it. It was recorded among the spark chamber tracks that back up the emulsion. The "V" was produced in the same microsecond as the charmed event and is almost certainly associated with the decay of the charmed particle.

25 November 1976

Monitor

European particle physicists have identified a new piece in the jigsaw puzzle of our knowledge of nuclear matter. One of 400 interactions of neutrinos observed in the Big European Bubble Chamber (BEBC) at CERN, Europe's centre for nuclear research, reveals a new particle, labelled the Σ_c^+. The newcomer is, like the proton, a baryon – in other words it contains three quarks, the entities that physicists believe comprise all nuclear matter. But this baryon is relatively unusual in that one of its quarks carries the property known as "charm".

The new particle is only the third charmed baryon to be found, and the first to be discovered in Europe by a team that includes members of Birmingham University, University College, London, and the SRC's Rutherford Laboratory.

The discovery is a "first" for the new track-sensitive target installed in BEBC. This allows physicists to study interactions with single protons in three cubic metres of liquid hydrogen, by looking at the tracks that the emerging particles produce both in the hydrogen and in a surrounding mixture of hydrogen and neon.

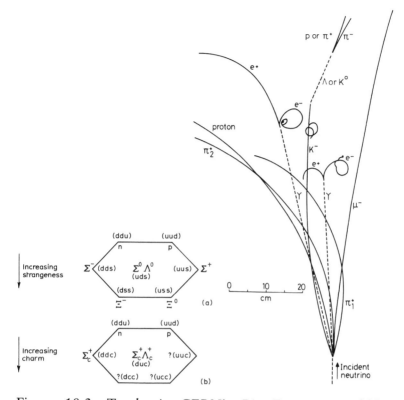

Figure 10.3 *Tracks in CERN's Big European Bubble Chamber (BEBC) seem to come from the decay of a heavier particle – in this case probably a charmed baryon, the Σ_c^+. This observation in 1980 filled a previously empty place in the pattern of baryons with spin-½, which includes the familiar proton and neutron. (Credit: RAL)*

The neon reveals the presence of gamma rays, by converting them to electron–positron pairs; this enables the physicists to infer the presence of neutral pi-mesons, as these particles decay into pairs of gamma rays. This identification is crucial in working back to find the mass of the particle created in the initial interaction between the neutrino and the proton (Figure 10.3) and to identify it as a Σ_c^+. The hydrogen "target" eliminates the confusion that would arise from the successive interactions with the many protons and neutrons that comprise a neon nucleus.

19 June 1980

Monitor

An experiment at the Stanford Linear Accelerator Center (SLAC) in California has cast new light on the decays of charmed particles, ephemeral subatomic particles that carry a property known as charm. The new results suggest that the lifetimes of electrically-charged charmed particles are similar to those of their neutral partners, contrary to earlier evidence.

Most types of sub-atomic particle are generally believed to consist of more basic entities called quarks, which come in only five or perhaps six varieties. One kind of quark possesses charm,

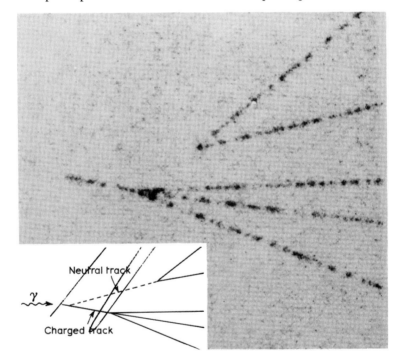

Plate 10.1 *The decays of neutral and positively charged charmed particles show up in this high-resolution picture from a bubble chamber at SLAC. A photon (γ) collides with a proton in the hydrogen contents of the chamber and produces the two charmed particles – one neutral, one positive – which live for a mere 6–8 × 10⁻¹³ s. The picture covers a distance of about 5 mm. (Credit: SLAC)*

which it then endows on any larger particle of which it is a constituent. Such charmed particles have so far proved difficult to study because they live for only $10^{-13} - 10^{-12}$ s, before breaking up, or decaying into more stable particles.

The experiment at SLAC, performed by an international collaboration of scientists from 18 institutions, has produced some of the best pictures yet of the decay of charmed particles. The particles are created in collisions between high-energy photons and protons, in the liquid hydrogen contents of a bubble chamber. The photon beam is produced when laser light scatters backwards off high-energy electrons from SLAC's linear accelerator. An advantage of this experiment is that it is performed close to "threshold", that is just above the energy needed to produce charmed particles, so the interactions are "clean" with little else being created.

Detectors outside the bubble chamber, "downstream" of the photon–proton collisions, provide some information about the particles produced and quickly determine, in less than 200 μs, whether or not the tracks of particles within the chamber should be photographed, so as to select the appropriate type of inter-action. An electronic flash is then activated and the camera photographs the trails of bubbles only 55 μm in diameter which the particles have left in the liquid hydrogen. This high resolution is necessary for the scientists to pick out the production and subsequent decay of the short-lived charmed particles.

The experimenters found decays of 11 neutral and 9 charged charmed particles, out of some 205 000 interactions. Their measurements indicate a lifetime of 8.2×10^{-13} s for the charged particles, and 6.7×10^{-13} s for the neutral particles. According to these results the lifetimes are similar, contrary to some earlier data, which suggested the lifetime of the neutral particle to be by far the shorter of the two. But all experiments still have large errors.

18 March 1982

11

The fifth quark revealed

Just when the world had come to accept four types of quark, a fifth joins the clan.

Leon Lederman and his group have certainly discovered a new kind of quark, if one can judge by the mood at the recent particle physics conference in Hamburg. Lederman's first tentative evidence for the existence of a particle at a mass of 10 GeV (ten times the mass of a proton) suggested that the particle might consist of a new quark bound tightly to its anti-particle. Since then the group has improved on its data, sharpened the peaks that represented the new state (so that its existence is now almost certain) and clearly resolved it into two or perhaps three particles. The resolution of the peak into separate parts enhances the probability that the particles represent new quarks.

The new state, which Lederman calls the upsilon, is likely to be equivalent to the J/psi, discovered at 3.1 GeV in November 1974, and its excited state.

The discovery of the J/psi was of great importance, for it was found to consist of a heavy charmed quark bound to its anti-particle. The existence of the charmed quark brought the number of quark types (or "flavours") to four, represented by the up, down, strange and charmed quarks. The existence of charm had been predicted by a powerful set of theories of the fundamental forces, the gauge theories, and this discovery established those theories and gave particle physicists a much-needed clue that they were on the right track.

So what need for a fifth quark (which will probably be called "beauty" or contrastingly "bottom")? Little need at all. But nature is full of surprises. Certainly few theorists want five quarks and if the upsilon does contain a fifth quark a search will immediately be

Plate 11.1 *Leon Lederman, now director of Fermilab, but in 1977 leader of the team that discovered the upsilon, a particle requiring the existence of a fifth type of quark.* (Credit: Fermilab)

on for the sixth ("truth" or "top"). Odd numbers of quarks are not easily fitted in to gauge theory models of the forces.

However even if we do not know how to predict the number of quarks — or their masses — the upsilon will turn out to be an excellent laboratory for investigating the strong force, one of the four fundamental forces and the one most difficult to handle theoretically or experimentally. The problem with the strong force is that it is so strong. The quarks are bound together by it into particles called "hadrons" (the proton and neutron are hadrons) and hadrons are then bound together by what's left over of the strong force (as in the nuclei and atoms). The effect of the force is overwhelming, and it is difficult to disentangle the stages by which the strong force takes effect. So it is difficult to test the modern strong force theory – the gauge theory called "quantum chromodynamics" in analogy to the quantum theory of electromagnetism, "quantum electrodynamics".

But at short distances and high energies the strong force becomes weaker and simpler. The forces between the heavy charmed quarks deep in the J/psi are already much simpler than the forces between light quarks in, say, the pion (which has similar constitution). The agreement between theory and experiment in predicting the excited states (internal states of motion) of the J/psi is one of the main reasons people now believe in quarks as real physical objects rather than theorists' whimsies. But there are still problems, and the upsilon, three times heavier than J/psi, will help to solve them.

MONITOR
8 September 1977

Monitor

Two experimental teams at the German colliding-beam accelerator in Hamburg (DORIS) have made the most accurate measurements yet on the new massive long-lived particle, the upsilon (Y). They have pinpointed its mass to 9.46 (±0.01) GeV, a resolution 50 times better than previously obtained.

The Y is intriguing because it implies the existence of a fifth type of quark. Just as the unexpected stability of the J/ψ could be explained satisfactorily only by its construction from charmed quarks, so the Y's longevity implies another new quark.

And the Y is very long-lived. This can be inferred from the sharpness of its mass by the uncertainty principle, which relates a short life to a large uncertainty in mass, and vice versa. The new mass of the Y is more than a 100 times sharper than one should expect at this energy, which means it lives 100 times longer. In fact, even this quoted uncertainty is almost entirely experimental, due to a lack of precision in the measurement of the accelerator beam's energy – so the Y is intrinsically even sharper than the figures show!

Physicists have already dubbed the new quark the bottom quark (and hope this is not all a midsummer night's dream, and that they will find another quark, the top quark at still higher energy).

The J/ψ was discovered simultaneously at an electron–positron machine (SPEAR) and at a proton accelerator (Brookhaven), but most of the subsequent precision work was done at SPEAR. This is because an electron and a positron colliding head-on, and interacting through only the electromagnetic force, is a much "cleaner" experimental test-bed than firing protons into a chunk of iron.

The Y was discovered at a proton machine, the giant Fermilab accelerator near Chicago, but its mass was too high for it to be produced by any of the existing positron–electron machines.

DORIS, however, was designed with some capacity in reserve – while rated at a centre of mass energy of 7 GeV, by pushing its components to their limits, the engineers there reckoned that DORIS could be coaxed to an energy of about 10 GeV. The Y was just reachable.

Instead of using many bunches of electrons and positrons in separate pipes, only one bunch of each was channelled into a single pipe. A final energy boost was provided by new units stolen, temporarily, from big sister PETRA, a new, more powerful, device now being built at the DESY centre in Hamburg.

In April this year, in a remarkably short time, the target was reached, a narrow bump in the data indicated the production of the Y – an amazing engineering feat. Immediately the power of the electron–positron device became apparent, for, by a roundabout method, the DORIS teams were able to estimate how likely the Y was to decay via the electromagnetic interaction as opposed to its more common decay by the strong force, a number not previously available.

This was an especially gratifying result for theoreticians, for by extrapolating standard models from the J/ψ mass to the Y, they can use it to tell whether the Y is made of a charge $\frac{1}{3}$ particle and its anti-particle, or out of charge $\frac{2}{3}$ particles. The Y fits the charge $\frac{1}{3}$ prescription, just right for the bottom quark.

Smashing a positron and an electron together is normally visualised as producing two excited quarks which separate at high speed. Subsequent interactions convert the quarks into normal nuclear particles, for *free* quarks have never been detected. But the initial conditions are "remembered" because two slews of nuclear particles, oppositely directed, are observed. The production of these "jets" has been recognised at the Stanford Linear Accelerator

Center and elsewhere, and theoretical analysis of their angular distribution shows that they could be produced by spin-½ entities – quarks.

The very stability of the Y indicates that such a prompt decay into two energetic quarks is unlikely at this precise energy. The quarks are quite happy sitting together – so how does the Y eventually decay?

One can draw an analogy here with orthopositronium, a bound state of an electron and a positron orbiting about each other. This decays through the electromagnetic interaction to produce three photons.

The analogy demands, therefore, that the Y should decay, by the strong force this time, and three gluons should be produced, these being, theoretically, the particles responsible for the strong force, in the same way that the exchange of photons is deemed to be the mechanism behind the electromagnetic force (see Chapter 18).

Gluons have never been seen, and are expected to be unstable; like quarks they will change before reaching the detectors. But just as with a quark jet, their original character would be revealed by the observing of three separate blobs of particles – gluon jets.

If one looks at the data from DORIS with a suitably Nelsonian eye it is claimed that there is some evidence for three jets. The more hardened would say merely that the events do not look like quark jets. Needless to say, no-one wants to go on record with such an important discovery until they are very, very sure, and certainty can come only with the collection of more data. The analysis will be difficult, but if three gluons are initially formed, their trajectories would lie in a plane. The final multiplicity of particles would preserve this by having a flatter, more pancake-like distribution than normal.

6 July 1978

Monitor

The discovery of a new particle by physicists in the US stengthens the belief that there are at least five types of quark – the building blocks of nuclear matter.

Experiments at the Fermi National Accelerator Laboratory near Chicago first revealed the *upsilons* – particles whose properties could be explained only in terms of a new type of quark which was labelled "bottom", or alternatively, "beauty". The three types of upsilon particle then observed all appeared to comprise a bottom quark bound with its anti-matter equivalent, but in slightly different configurations so as to give rise to particles with three different masses. Now a team at the Cornell Electron Storage Ring (CESR) in Ithaca, New York, has observed particles that appear to contain a bottom quark, but bound this time with a different variety of quark.

There are now known to be at least four types of upsilon particle, and they are the heaviest particles yet found, with masses in the range 9.4–10.5 GeV, roughly 10 times the mass of the proton. The heaviest of these was first reported earlier in 1980 by two groups working at the CESR machine, which is ideally suited to producing upsilons in collisions between beams of electrons and positrons (*Physical Review Letters*, vol. 45, p. 219 and p. 222).

The fourth upsilon turned out to be different from its lighter relatives in that it apparently decayed to other particles much faster. The first three upsilons decay relatively slowly, implying that the bottom quark and anti-quark have difficulty getting away from each other to form other particles. The particles probably decay instead by first producing three gluons (see Chapter 16). The faster decay of the fourth upsilon suggests that it has simpler alternative decay mechanisms.

The team working on the detector called CLEO at CESR has studied the decay products of the fourth type of upsilon and has found good evidence for B-mesons – particles made of a bottom quark bound with another type, probably one of the varieties that make up protons and neutrons. The researchers found the B-mesons by looking for electrons and muons in their detector, as conventional theory says that B-mesons should decay readily only via the weak nuclear interaction, and electrons and muons can come only from such decays. Future work should determine the nature of the other quark in the B-mesons.

11 September 1980

Monitor

Physicists working on CESR, the Cornell Electron Storage Ring at Cornell University in Ithaca, New York, have found new evidence for the so-called B-mesons. These are subatomic particles that are similar to the more familiar pi-mesons, but which are made from a different combination of quarks, the fundamental building blocks of matter. Although the same group of physicists has found good evidence for the production of B-mesons at CESR before, this latest result shows for the first time the explicit decay of B-mesons into less-rare, longer-lived particles (*Physical Review Letters,* vol. 50, p. 881).

The protons and neutrons of everyday matter appear to be built from two types of quark, called "up" and "down" or *u* and *d*. But short-lived particles produced in experiments at high-energy particle accelerators turn out to contain other types of quark, known as "strange", "charmed" and "bottom", or *s, c* and *b*. The B-mesons contain only one *b* quark (or anti-quark) bound instead to a more "ordinary" *u* or *d* anti-quark (or quark).

It turns out that the heaviest of the four upsilon particles has sufficient mass to decay radioactively into two B-mesons; the *b*

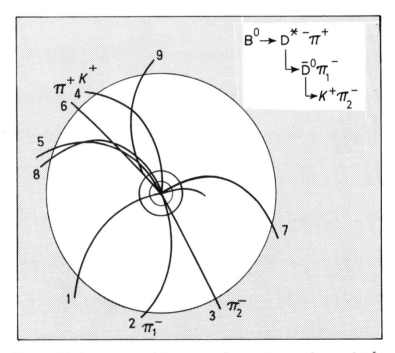

Figure 11.1   *A neutral B-meson decays into a charmed D***-meson and a pi-meson (π⁺), eventually to produce a K-meson and three πs, all of which produce tracks in the CLEO detector at the Cornell Electron Storage Ring (CESR). The B-meson, which contains a bottom quark bound with a more common up or down quark, was produced in a head-on collision between beams of electrons and positrons.*

quark and anti-quark that it contains split up to go their separate ways in the two B-mesons. CESR is the only machine that can produce reasonably large numbers of the heaviest upsilon, so its collisions between electrons and positrons provide the most promising hunting ground for B-mesons.

This time the researchers have looked for steps in the decay chain of a B-meson, in which the *b* quark converts first into a *c* quark and then into an *s* quark; this means that the B-meson decays into a charmed D-meson, and then a strange K-meson. The K-mesons live long enough to produce tracks in the CLEO detector, so the team could work back from detected K-mesons to hunt for those that seemed likely to have originated from a B-meson. In such cases the spray of particles emanating from the electron–positron collision would be entirely consistent with the steps expected for the decay of a B-meson (see Figure 11.1).

From the 18 possible B-mesons they were able to find, the researchers have deduced a mass of 5270 MeV for the neutral B-meson and 5270 MeV for the charged B – some 5.3 times as heavy as the proton. The B-mesons can now be assured of a place in the tables of known sub-atomic particles.

14 April 1983

Monitor

Two teams of physicists in the US have measured for the first time the brief life of the sub-atomic particle known as the B-meson. The result provides a possible means of testing what has become known as the "standard model" of sub-atomic particles and their interactions.

The B-meson has appeared only recently in experiments in which electrons and positrons (anti-electrons) colliding at high energies annihilate each other. Their combined energy can rematerialise as particles in a variety of ways, provided the basic laws of physics are obeyed. Many of the particles produced are called "hadrons", particles like the proton and neutron, which appear to be built from more basic entities called quarks. The quarks, it seems, cannot appear alone, but those formed in the electron–positron annihilations materialise in clusters that form the particles we can observe.

The B-meson is unusual in that it comprises the so-called "bottom" quark (*b*) – the heaviest known quantity – along with another lighter quark. (Strictly speaking the meson is a combination of a quark and an anti-quark.) Being over five times as heavy as the proton, the B-meson can decay to a lighter particle but only if a *b* quark changes to a lighter quark, a reaction that is possible only via the weak nuclear force. The lifetime of the B-meson is therefore relatively long in sub-atomic terms, but exasperatingly short when it comes to attempts to measure it in the laboratory.

The two experiments are at the Stanford Linear Accelerator Center in California, where there is an electron–positron collider called PEP. Both teams pinpoint B-mesons by searching for high-energy electrons and muons emitted in the decays. These particles are quite distinct from the many hadrons produced in the annihilations and so provide a relatively clean "signature" for the decay. The appropriate electrons and muons are further distinguished because they leave the site of the decays at relatively large angles.

It then remains to track the particles back to estimate how far from the collision point each decay occurred; this distance is related to the time the B-meson lived. The distances involved are so small – a few hundred micrometres – that only a statistical average over many decays is possible. But the two teams, working on the detectors known as MAC and Mark II, have succeeded in calculating the lifetime of the B-meson. The results they have are 1.8×10^{-12} s and 1.2×10^{-12} s, respectively, with errors of around $\pm 0.5 \times 10^{-12}$ s.

The "standard model" of particles and their interactions suggests that there should be six quarks rather than the five that form all the particles observed so far. The model intimately links the bottom quark with a "top" quark (t), for which there is as yet no firm evidence, and relates the mass of the top quark to the lifetime of the B-mesons. Researchers at CERN, Europe's centre for particle physics near Geneva, are searching for signs of the top quark. Once discovered, the t and b quarks will test neatly the standard model.

1 September 1983

12

A new lepton

MARTIN PERL

An experiment reveals a heavy lepton – a new relation for the electron and its mysterious sibling, the muon.

The existence of a new particle, similar to the electron and the muon but much heavier, has become considerably more likely with the latest results from SPEAR, the electron–positron storage ring at Stanford, California. Physicists there have used a new apparatus and found a signal "compatible with the expected decays of pairs of heavy leptons". As nature presently appears to be based on the existence of just four leptons (point-like particles with no strong interaction) and four quarks (point-like particles with strong interaction) another lepton would be of some consequence.

More than a year ago Martin Perl, a team leader at SPEAR and an independent spirit, claimed he had found events which showed the creation and decay of the possible new particle (which he called U for "unknown"). Until this week the latest data were in a paper describing 105 events published last year (*Physics Letters*, vol. 63B, p. 466). The events showed the production, after the collision of an electron and a positron in SPEAR, of a pair of oppositely charged particles, one an electron and one a muon, associated with no other detectable particles. The events occurred only when the total energy of the collision was about 4 GeV.

The group at SPEAR considered that these results were "consistent with" the occasional production of a pair of new particles, U^+ and U^-. The particles then decayed, analogously to the decay of the muon, by producing an electron and two neutrinos or (as the particle would be heavier than the muon) a muon and two neutrinos. The muon and the electron would be the only particles detected. Although pairs of electrons and pairs of muons would be

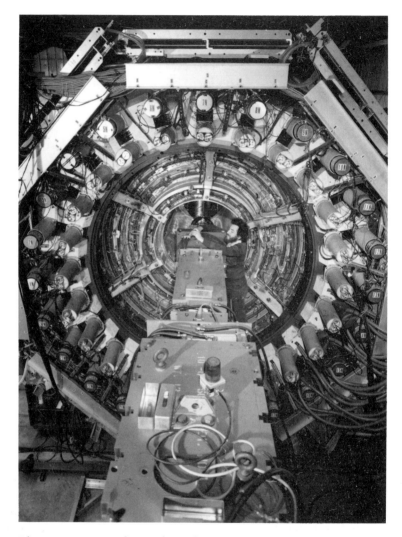

Plate 12.1 *Right: The electron–positron storage ring, SPEAR, at SLAC is to the right of one of the end stations fed by the linear accelerator. Here the team working on the Mark I detector (above), discovered the new lepton, the tau. This is the same machine and detector that co-discovered the J/psi and the family of charmed particles. (Credit: SLAC; Lawrence Berkeley Laboratory)*

equally expected (at half the frequency) their existence would be totally masked by normal electrodynamic processes and were thus not searched for.

The new results, announced in *Physical Review Letters*, vol. 38, p. 117, are based on an improvement in the efficiency with which the SPEAR magnetic detector sees muons – at least in one direction. Concrete absorbers filter out other particles but let muons through to be detected in the spark chambers beyond. The group looked for events in which one muon was produced, plus anything else, in three energy ranges. They found a large "anomalous muon" signal, meaning muon events which had no conventional explanation, in each energy range.

For events with just two charged particles including the muon (the number expected from the decay of a pair of heavy leptons) the size of the anomaly neatly matched the production and decay of heavy leptons of a mass between 1.6 and 2.0 GeV. This was true in each energy region. So the likelihood is now that a new charged partner for the electron really exists.

MONITOR
20 January 1977

Leptons – what are they?

The history of science shows that one of the best ways to make progress in a scientific field is to study the simplest objects or systems in that field. However, this has not been the case in elementary particle physics, where the simplest objects are the electron, the muon and the neutrinos – the particles we call leptons. The study of the leptons themselves has led to more questions than answers, and we have learned more about the basic nature of matter by studying much more complicated particles such as protons, or by studying the interactions of leptons with these more complicated particles. This may now change with the discovery in the past few years of a new and heavy charged lepton called the tau. We may be on the verge of answering a few of the questions raised by the electron and the muon. Incidentally, the name lepton comes from the Greek prefix *lepto* meaning small or slight. This name was devised in the belief that these particles have small masses compared with other particles. The discovery of the tau with its large mass destroyed this belief, but the name remains.

The electron (*e*) was the first elementary particle to be identified; J. J. Thomson discovered it in the 1890s. All atoms consist of electrons orbiting about a central nucleus, so it is very easy to study the properties of electrons in atoms. It is also easy to remove electrons from atoms and to study them in isolation, free from the influences of all other matter. Thus the electron has been studied more than any other elementary particle. It has an electric charge of 1.6×10^{-19} coulombs which can be either negative or positive. (When it is positive we call the particle a positron.) Its mass is 9.1×10^{-28} g, but this is usually expressed in equivalent energy units MeV (million electronvolts). It is then a more easily remembered number, namely 0.51 MeV.

The electron's size is less than 10^{-15} cm, which is less than 1/100 the size of a nucleus and less than 1/10 000 000 the size of an atom. Indeed the electron seems to be simply a point particle with no internal structure to the best of our knowledge. Unlike the proton which has quarks inside it, the electron does not have other particles inside it. There is another fundamental difference between the electron and the proton; protons, and neutrons also, are acted upon by a strong, attractive force which holds them together and thereby holds the nucleus together. This "strong force" does *not* act on electrons at all! This is another aspect of the electron's simplicity, for the strong force causes particles to act in very complicated ways.

Electrons, being electrically charged, are acted upon by the electromagnetic force. They are also acted upon by the "weak force" and by the gravitational force. The weak force is weaker than the electromagnetic force in most reactions, hence its name, but it is stronger than the gravitational force. However, the weak force acts only across short distances (10^{-13} cm or less) so we do not observe this force in everyday life.

This is most of what we know about the electron – except for how it interacts with neutrinos and other charged leptons – a subject I shall come to later. It is indeed a simple particle, yet it leaves us with questions to which we have no answer. What sets the electron's charge at 1.6×10^{-19} coulombs? Why is its mass 0.51 MeV and not 5 MeV? And there is an old question first pondered by Hendrik Antoon Lorentz in the early 1900s, which remains unanswered. The negative charges "in" the electron all repel each other, so what then holds the electron together?

In the late 1930s the second charged lepton, the muon (μ) was discovered in cosmic rays. Many of the properties of the muon are the same as those of the electron (see Table 12.1): the muon's size is less than 10^{-15} cm; the muon has no internal structure; the muon is not acted upon by the strong force, but is acted upon by the electromagnetic, weak and gravitational forces; its electric charge is the same as that of the electron. But the muon's mass is about 207 times the mass of an electron. Thus the existence of the muon raises yet another question. How can two particles alike in so many ways, differ so greatly in mass?

Because there are positively and negatively charged electrons and muons, it is natural to inquire as to the existence of electrically neutral electrons or muons. The latter particles do not exist. However, the electron and muon do have electrically neutral partners, the electron neutrino (v_e) and the muon neutrino (v_μ) respectively. These neutrinos have zero or close to zero mass (Table 12.1), unlike the electron or muon. We do not know why this is so. Neutrinos, like electrons and muons, are not acted upon by the strong force, but because they are electrically neutral, neither are they acted upon by the electromagnetic force. They are acted upon by the weak force, however, and, if they possess energy, by the gravitational force. As far as we can tell, neutrinos are stable particles.

At this point the sceptical reader might ask, "The muon is

Table 12.1   *The known leptons and their properties*

Charged lepton name	Electron	Muon	Tau
Charged lepton symbol	$e^{\pm}$	$\mu^{\pm}$	$\tau^{\pm}$
Charged lepton mass (MeV)	0.51	105.7	about 1785
Charged lepton lifetime (seconds)	stable	2.2×10^{-6}	less than 3×10^{-12}
Associated neutrino symbol	$\nu_e, \bar{\nu}_e$	$\nu_\mu, \bar{\nu}_\mu$	$\nu_\tau, \bar{\nu}_\tau$
Associated neutrino mass (MeV)	less than 0.00006 (may be zero)	less than 0.57 (may be zero)	less than 250 (may be zero)
Ratio of lepton mass to electron mass	1	207	3510

differentiated from the electron by its larger mass; but if the muon neutrino and electron neutrino both have no mass, what differentiates them?" Our unsatisfactory answer is that we don't know what differentiates them in a basic way, but experimentally we find the following. An electron neutrino can make electrons or other electron neutrinos when it interacts with matter but it *never* makes muons or muon neutrinos. Conversely a muon neutrino makes muons or other muon neutrinos but never makes electrons or electron neutrinos. An electron neutrino has a basic "electronness" quality and a muon neutrino has a basic "muonness" quality, and these are completely different.

These "electronnes" and "muonness" qualities are also exhibited by the electron and muon themselves. For example, although the muon is unstable and although its mass is much greater than the electron mass, a negative muon never decays simply to a negative electron plus electromagnetic energy. The decay proceeds through the weak force in a much more complicated way, producing a muon neutrino and an anti-electron neutrino (ν_e) along with an electron. We explain this by saying that the "muonness" quality of the muon cannot convert into the "electronness" quality of the electron. Instead the "muonness" of the muon has to be carried off by the "muonness" of the muon neutrino and the creation of an electron with its "electronness" quality is compensated for by the simultaneous creation of an anti-electron neutrino. If all this sounds like tautological nonsense, in a sense it is. As I said earlier, we don't know what differentiates in a basic way a μ from an e or a ν_μ from a ν_e. This was the state of our understanding about four years ago when I and my colleagues from the Stanford Linear Accelerator Center and the Lawrence Berkeley Laboratory began a search for other charged leptons. I was motivated by an old idea in scientific work; if you can't understand a set of objects, try to find more examples of such objects. We hoped to produce new charged leptons using a new facility at Stanford called SPEAR, in which electrons and positrons collide at high energy. The electrons and the positrons annihilate into a very small bundle of electro-magnetic energy, which then produces a pair of new charged leptons (Figure 12.1). "Hoped" is probably too strong a word to use because we didn't know if charged leptons other than electrons or muons existed, and if they did exist we didn't know if our experiment had enough energy to make new leptons. Thus it was to our surprise, and I think to the greater surprise of other elementary particle physicists, that we found a new charged lepton called the tau (τ). Other experimental groups at SPEAR and at a similar electron–positron facility at the Deutsches Elektronen-Synchrotron, DORIS, later helped to confirm this discovery.

I don't want to take the space here to explain in detail the evidence for the tau being a lepton. Briefly, the properties of the tau are analogous to those of the electron and muon. It is a simple particle with no internal structure; it is acted upon by the electromagnetic, weak and gravitational forces, but *not* by the strong force; it has the same size electric charge. All existing measurements are in agreement with the tau's being a lepton. However, since the tau is newly discovered we must continue to test its leptonic nature.

The most astonishing property of the tau is its mass of about 1785 MeV. We call it a "heavy" lepton, the term heavy having two connotations. First, the tau is about 17 times heavier than the muon and 3500 times heavier than the electron. Secondly, the tau is heavier than many non-leptonic particles. For example, its mass is about twice that of the proton or neutron, and it is about 13 times heavier than the pion, the particle which carries the strong force and hold the nucleus together.

The significance of this large mass is first that it destroys the once popular belief that the leptons had to have small masses. (Indeed the name heavy lepton is paradoxical.) Secondly and more

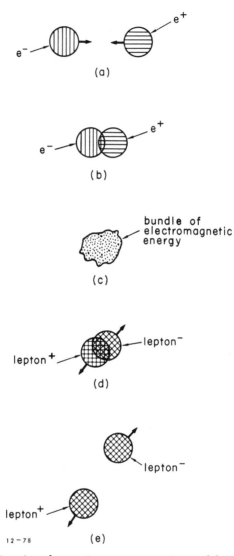

Figure 12.1 *A schematic representation of how electrons (e⁻) and positrons (e⁺) collide to make new leptons. The e⁺ and e⁻ moving toward each other in (a) collide in (b) to annihilate and form a bundle of electromagnetic energy in (c). This bundle of energy converts into a pair of leptons in (d) which separate and move away from each other in (e).*

significantly, now that one heavy charged lepton has been found, it is reasonable to expect that more and even heavier charged leptons exist. We know of no general physical principle which puts an upper limit on the mass allowed for a heavy charged lepton, or which limits the number of different kinds of such leptons. Just as the electron and muon possess mysterious and unique qualities which I have called "electronness" and "muonness", so the tau possesses a unique quality which I call "tauness". This has been demonstrated experimentally in two ways. First, the tau *cannot* be made by the interaction of muon neutrinos with matter. Secondly, when the tau decays to electrons or muons it goes through a process analogous to that for muon decay, producing a tau neutrino to carry off the "tauness" of the tau.

We know less about the tau neutrino than we do about the other neutrinos because all this is so new. We believe the tau neutrino is uniquely associated with the tau but the experimental tests of this belief are much less extensive than are the tests that prove that the electron neutrino is uniquely associated with the electron. Also, as shown in the Table 12.1, the mass of the tau neutrino is smaller than the mass of the tau but we still have a long way to go to prove that the tau neutrino has a mass as small as that of the other neutrinos.

I began this article by writing that I thought the discovery of the tau would help to answer a few of the questions raised by the existence of the leptons in the first place. Let me indicate how this might happen. I don't think the tau will help us to answer questions such as: how can we calculate from basic principles the magnitude of the electric charge (1.6×10^{-19} coulombs) possessed by all the charged leptons? This question, like the question of how to calculate the speed of light from basic principles, is so intractable that it must be left for future generations of hopefully brighter physicists. The sort of question that the tau might help us to answer is: what is the equation or rule that sets the masses for the charged leptons? I hope that we are beginning to see a series of masses analogous to the role of spectra in the development of atomic physics. Just as optical spectra were the key to our obtaining a fundamental understanding of atomic physics, so might the charged lepton mass spectra be the key to a fundamental understanding of lepton physics.

22 February 1979

13

The generation game

FRANK CLOSE

The growing numbers of elementary quarks and leptons point to a curious redundancy in nature, which may be allied to a fundamental symmetry of the Universe.

During the past 30 years physicists have built machines that can accelerate protons to very high energies and then smash them into a target. Upon impact the protons shatter the nuclei of the target's atoms, and physicists have found among the debris many new types of particle which survive for only a minute fraction of a second. These particles all experience the strong attractive force that fused the protons and neutrons together in the nuclei in the target, and so they appear to have as much claim to be called fundamental as do the neutron and proton.

But, on aesthetic grounds there seemed too many particle varieties for them all to be fundamental. From 1945 to 1974 physicists found over 100 examples of particles which feel the strong nuclear force (called collectively hadrons), while they discovered only two new examples of leptons (particles, like the electron, which do not respond to the strong force). These new leptons were the muon – seemingly a heavier version of the electron – and a second neutrino. The discovery of just two more leptons in the search for elementary particles seems reasonable, but finding more than 100 hadrons is a worry – an embarrassment of riches. As a result physicists focused attention on the strong force and the hadrons: understand why there were so many hadrons and, they suspected, they could make major progress.

In the course of studying the strong force, patterns of regularity appeared among the hadrons. These patterns, known as the "eightfold way", turned out to play a seminal role in the subsequent development of the physicists' understanding of elementary particles. Just as Mendeleev's observation of atomic regularities was later understood to be a consequence of the atom's substructure of "electron-plus-nucleus", so the "eightfold way" is now understood to be a manifestation of substructure in the hadrons: they are clusters of subatomic entities called "quarks".

Physicists today believe that this level of matter contains the true building blocks: the elementary particles are leptons and quarks. I will temporarily defer answering the question "how many?"

The vast majority of the hadrons can be understood by postulating the existence of just two types, or "flavours", of quark; an "up" quark (*u*) and a "down" quark (*d*). The neutron and proton, for example, are *ddu* and *uud* respectively. The quarks have the property that their electrical charges are two-thirds of the proton's charge for the *u* quark, and minus one-third of the proton's charge for the *d* quark – a feature unique to quarks.

The above is nearly – but not quite – the whole quark story. Hadrons found before 1974 that did not fit in with the "up-down" quark picture also happened to share an anomalous property known as "strangeness". Physicists could account precisely for all these "strange" particles at the expense of postulating the existence of a third flavour of quark, called the "strange" quark (*s*). So some semblance of order had returned; the world appeared to be built from four leptons (electron and its neutrino, muon and its neutrino) and three quarks (up, down and strange).

This was the situation until 1974, when another unusual particle, the J/psi meson, was discovered, and found to live some 1000 times longer than current conventional theories expected. The reason for the J/psi's longevity that has come to be accepted is that it contains a quark and an anti-quark, each endowed with equal but opposite values of yet another quark flavour, "charm" – a name coined by theorists Sheldon Glashow, now at Harvard University, and James Bjorken at Stanford, who required a fourth quark flavour in their theory of quark–lepton symmetry. The J/psi thus carries "hidden" charm – while having zero charm itself, it does contain charmed quarks, which find it difficult to transmute to quarks of more "ordinary" flavours. However, since the discovery of the J/psi, experimenters have also found particles that do carry "naked" charm.

One of the great successes of the quark hypothesis was that physicists used it to predict that there must exist in nature a type of matter that had been hitherto unknown, possessing a property called "charm". Since 1974 considerable evidence has accumulated confirming the existence of this new type of matter, and the existence of a fourth flavour of quark – a "charmed" quark (c).

Charmed matter and strange matter (containing one or more charmed and strange quarks respectively) can transmute into one another and produce a muon and neutrino as by-products. This is analogous to the way a neutron transmutes into a proton – a process that is really only a "down" quark transmuting to an "up" – with electron and neutrino produced (see Box 13.1 p. 135). In fact this similarity appears to be fundamental. It is as if in the charm and strange quarks together with the muon and its neutrino, nature has made Xerox copies of the up and down quarks with the electron and its neutrino. These quark and lepton pairs are referred to as "generations" of elementary particles; the up and down quarks, electron and its neutrino being the first generation; the charm and strange quarks, muon and its (second) neutrino being the second generation. And all phenomena involving the first generation appear to have an analogue in the second generation.

Why nature "repeats" itself, physicists do not yet know. The familiar world about us appears to survive quite well with only one pair of quarks (up and down) and one pair of leptons (electron and neutrino). These particles alone are sufficient to explain the structure of the atom and the interaction between different atoms – the basis of chemistry. But the existence of the second generation of particles seems to be an extravagance, and does not appear to affect our everyday experience in any significant way. We do not yet understand why the second generation is there at all.

In 1975 Martin Perl and his colleagues at Stanford in California discovered a third generation of leptons (see Chapter 12). These new leptons are the "tau", and its neutrino. The tau is more massive than the proton, so the name "lepton", originally meaning "light", is something of a misnomer – it is for this reason that we now use "lepton" to denote particles that do not feel the strong force. Is there a corresponding third generation of quarks? This is one of the major questions that physicists are currently investigating. A fifth quark, "bottom" (b), with charge minus one-third, was discovered in 1977 and is a prime candidate for belonging to a third generation of quarks. If there is indeed a third generation of quarks then a "top" quark with charge two-thirds, associated with the b and its minus one-third charge, is waiting to be discovered. There exist several predictions for the mass of the lightest hadrons containing a "top" quark, but they must be some 35 times that of the proton to have avoided detection. As an experiment must produce such new hadrons two at a time — a "top" quark and a "top" anti-quark must be produced together because the everyday world does not contain "topness" — then they may escape detection for some years. If discovered, the new hadrons would put the seal on the idea that nature has created several generations of quarks and leptons. [In July 1984 the UA1 experiment described in Chapter 22 found the first evidence for a top quark.]

If nature has indeed chosen to pair quarks and leptons this way then we are already being presented with a suggestion of a connection between them. In a sense we are already finding hints of a "latter-day Mendeleev Table". The cause of the pattern Mendeleev found was a deeper level of structure beyond the atom, while the patterns of the hadrons in the "eightfold way" were due to the quarks within. Are we therefore seeing in these patterns of generations of quarks and leptons the first hints of something beyond the quarks and leptons themselves, or is their relationship more profound? There is a widely-held belief that quarks and leptons are in some way related, but there is no consensus yet as to how.

Recent theoretical ideas postulate that at the time of the big bang all of nature's forces were unified; they then froze out into the gravitational, weak, electromagnetic and strong forces as the Universe cooled to its present form. High energy accelerators create very hot conditions in a small region of space which enable us to test these ideas and there are some hints that they may not be far from the truth. In particular these theories predict the strong force to be necessarily stronger than the electromagnetic force in the cool Universe. In some versions of these theories the masses of leptons and quarks are related to one another. In each generation the mass of the quark *d, s, b* is bigger than that of the corresponding charged lepton, but the ratio gets smaller as the mass increases. Table 13.1 shows that this indeed appears to be the case. If further generations are found and if the theory is correct then lepton and quark masses could approach one another, eventually becoming equal if masses of the order of 10^{15} times the proton mass are found(!). At this scale of mass, gravity enters the picture and no one is yet entirely sure what happens.

Understanding the menu of quarks and leptons is an important problem. Astrophysical calculations claim that there can be at most one more neutrino. This might imply at most one more generation, or that we have already found them all. If so, why

Table 13.1   *Ratio of quark mass to lepton mass*

1st generation	2nd generation	3rd generation
$\dfrac{d}{e^-} \gtrsim 20$	$\dfrac{s}{\mu^-} \simeq 5$	$\dfrac{b}{\tau^-} \simeq 2$

three (or four) generations? At the moment this is a mystery as deep as that of why nature was not satisfied with just one generation. Perhaps the very fact that we exist and are able to ask the question has required a delicate balance between the rates at which stars and life evolve that would not have resulted had only one generation existed. It is also possible that three generations may be necessary to understand the asymmetry in nature between forwards and backwards in time. But why this is a necessary rather than an amusing phenomenon is not known.

Why nature distinguishes between left and right in space, forwards and backwards in time and has in general developed the way that it has, are still metaphysical questions, but the recent discoveries are bringing us nearer to the point where these questions can be formulated scientifically. Insights into some or all of them may emerge as the view in the lepton–quark theatre becomes clearer.

29 November 1979

Box 13.1 Quark transmutation

Radioactive transmutation of a neutron through beta decay occurs when a down quark (d) becomes an up quark (u), emitting an electron and a neutrino (which is technically an *anti*-neutrino). A neutron is formed from two down quarks and one up quark, while a proton consists of one down and two up quarks. Consequently the act of converting a down quark into an up quark causes a neutron to become a proton (Figure (a)).

This act of transmutation involves all of the first generation of elementary particles, namely up and down quarks, electron and neutrino leptons (Figure (b)). The analogous process involving second generation particles is shown in Figure (c). Because the charmed quark is more massive than the strange quark, this second generation reaction tends to proceed in the opposite direction in

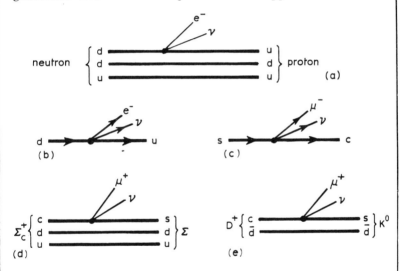

nature, that is $c \rightarrow s\mu^+\nu$. The μ^+ is the positively charged anti-particle of the muon. The direct analogue of the beta decay of a neutron is shown in Figure (d); experimenters have observed this reaction, as well as the decay of a charmed particle, a meson, formed from a quark (c) plus an anti-quark (d) (Figure (e)).

PART FOUR

The glue in the atom

Over the past 30 years or so, particle physicists have discovered a vast range of ephemeral subatomic particles, most of which can be classified as hadrons – particles built from quarks. These new particles are generally produced in high-energy collisions between more stable objects, such as protons, either in the laboratory or high above the Earth where the cosmic radiation enters the atmosphere. But where do the new particles come from?

The "easy" answer is that they materialise from the energy of the collisions, in accordance with Albert Einstein's equation, $E = mc^2$, where E is energy, m is mass and c is the velocity of light. The more energy available, the more particles can be created, and the heavier they can be. But this is not the whole story. If the particles are hadrons they must be created through the agency of the strong nuclear force, and it is only through understanding this force that we can learn how the hadrons are created, and destroyed.

The strong nuclear force was first identified 50 years ago, long before quarks were thought of, as the force that binds protons and neutrons within atomic nuclei. In 1935, Hideki Yukawa proposed that the strong force is transmitted by particles with a mass some 15 per cent that of the proton – the pi-mesons, or pions. The protons and neutrons (nucleons) within a nucleus are bound together as they exchange "virtual" pions. The pions are said to be "virtual", for their mass is created only temporarily: one nucleon emits a pion, another absorbs it. The process is governed by Werner Heisenberg's uncertainty principle. Generally speaking, the total energy of a system must always remain constant, but the uncertainty principle allows a variation in energy (and hence, through $E = mc^2$, a variation in mass) for a limited period of time; the larger the amount of energy (or mass) the shorter the period of time. The mass–energy of the pion is roughly equivalent to the time it takes light to cross a nucleus – the fastest that any sub-nuclear process can occur.

The pions inside a nucleus are virtual, coming and going constantly as they are emitted and absorbed by the nucleons. But suppose a nucleus is given some energy, equivalent, say, to the mass of a pion, or more. Then "real" pions can emerge from the nucleus, released by the additional energy from the "virtual" cloud of pions that carry the strong force. So when nuclei, or even single protons, are bombarderd by high-energy missiles such as other protons, pions are produced.

But we now know, as Parts Two and Three have revealed, that hadrons – including protons and pions – appear to be built from quarks. So where does this leave our picture of the strong nuclear force?

The ability to study quarks in a proton is like being able to observe the structure of the atom; in describing matter in terms of quarks we are describing it in more detail. Now, we can study the ways that atoms interact to form molecules without knowing of their detailed structure. In the same way, knowledge of quarks within the protons and neutrons was not necessary for Yukawa to describe the way that neutrons and protons bind together. And just as the origins of the molecular forces lie not in the neutral atoms but in the electric charge of the electrons within the atoms, so the strong nuclear force originates with the quarks, rather than with the protons or neutrons.

The present theory of the strong force is called *quantum chromodynamics* (QCD). The prefix "chromo" refers to a property, or "charge", unique to the quarks which is called "colour". James Dodd introduces the concept of colour and the basic ideas of QCD in Chapter 14. The theory is analogous in some respects to Yukawa's, but in QCD the quarks exchange entities called "gluons". In Chapter 15 Frank Close brings ideas on QCD up to date with a discussion of novel forms of "gluematter", including hybrids of quarks *and* gluons. And in Chapter 16 we see what experimental evidence there is for gluons, and for QCD.

The strong force must be very peculiar, for it seems, as we saw in Part Two, that experiments can never liberate quarks from within particles, however much energy we supply. Instead, such experiments produce pions – just as Yukawa envisaged – which are themselves built from quarks coupled with anti-quarks. We can also create baryons, provided there is enough energy to produce *anti*-baryons at the same time, to keep the overall balance of matter and anti-matter. Moreover, as Part Three revealed, collisions of protons can create charmed and bottom particles, as well as the strange particles first seen in cosmic-ray experiments. In other words we can create particles containing types of quark that apparently do not exist within the proton. Or do they?

According to QCD, a proton contains three quarks, which are bound together by exchanging gluons. But just as virtual pions are continually being created and destroyed in Yukawa's theory, so pairs of quarks and anti-quarks are created and destroyed in a particle like a proton. These quark–anti-quark pairs need not be of the same variety as the three "valence" quarks that are the basic constituents of the proton, and which give it its characteristics. Thus with sufficient energy, collisions can knock not only pions out from protons, but also strange particles, and with more energy still, the heavy charmed and bottom particles; but never single quarks.

QCD is indeed a surprising theory. Moreover, it proves very difficult to perform calculations within the theory, partly because the strong force is so very strong. In Chapter 17 David Wallace charts the latest attempts to harness the power of huge modern computers to perform QCD calculations, and further comprehend the "glue in the atom".

14

A theory for the strong force

JAMES DODD

To explain the behaviour of quarks and the strong force that binds them within particles, theorists find they must give the quarks novel properties, dubbed "flavour" and "colour".

Over the years, physics has opened up matter like a nest of Russian dolls. Everyday matter is made of molecules that consist of atoms bound loosely together in the familiar chemical configurations. The atoms are made of electrons bound in orbits around a central nucleus; the simplest atom, hydrogen, has just one electron and one proton. The atomic nuclei of all other elements consist of protons and neutrons bound together by the strong nuclear force. Physicists now believe that there is one more layer; that the neutron, proton, and other similar particles, referred to generically as the hadrons, are themselves composed of quarks, the fundamental building blocks of nuclei and, ultimately, of matter.

But quarks remains elusive. No matter how energetically the protons are made to collide into each other, no constituent particles emerge. This is surprising because the proton has a size we can measure, and high energy accelerator experiments have indicated its detailed substructure. Despite this, individual quarks remain unobserved; but in a sense such behaviour is welcome. If quarks really are true "elementary" particles, then we might expect them to display some unique characteristics as they interact through the fundamental forces in nature.

The strong nuclear force is responsible for the interactions of hadrons, and whatever might be inside them, for instance quarks. In its most familiar manifestation, the strong force binds together protons and neutrons inside the nucleus. To gain any fundamental understanding of this force we must discover the structure of the hadrons themselves, and this we can achieve only by experiments, in which particles collide into each other and, we hope, probe their own structure. In these experiments, microscopic particles travel close to, or even at, the speed of light, and their description requires a combination of the two great theories of modern physics. First we must use Einstein's special theory of relativity to be able to describe these interactions in which such high energies and velocities are involved. Secondly, we need quantum mechanics, the accepted theoretical description of microscopic phenomena.

One of the simplest observations in these experiments is that particles are often created or destroyed in the collisions, so the particles themselves are unsatisfactory elements on which to base a theory. Instead, physicists imagine "fields" extending through space and time, and they represent the particles as locally manifest energetic states, technically called "quantum excitations", of these underlying fields. To remind us that these fields are associated with particles we can call them "matter fields".

Another simple observation is that there is no net creation or destruction of electrical charge in the collisions regardless of what happens to the particles. But a simple description of interactions between particles, or the "matter fields", cannot automatically account for the conservation of electrical charge. It is necessary to add another field to the theory to cancel out unwanted reactions and this is called a "gauge" field. In quantum electrodynamics (QED) – the quantum theory of electromagnetic interactions – this gauge field is just the electromagnetic field, and it too has quantum excitations (in everyday terms, particles) called photons. The force between electrons is then described by the exchange of photons between them – a mechanism that can be likened to the way in which the exchange of a rugby ball causes players to "interact" as the ball is passed between them.

From these beginnings, physicists have built up a relativistic quantum theory of gauge fields, and it is with this powerful technique that theorists are now trying to describe the behaviour of quarks.

One of the most remarkable things about quarks, according to the standard theory, is that they carry a fraction of the charge, e, of the electron, once believed to be the fundamental unit of charge. To explain the occurrence in nature of positively charged, negatively charged and neutral particles, we need two kinds of quark called "up" and "down". The positively charged proton can be made from two "up" quarks (each with charge $+\frac{2}{3}e$) and one

Box 14.1 Quarks and quantum numbers

Six different types of quark can explain the large variety of all known particles, excluding the leptons – the electron, its relations the muon and the tau, and the neutrinos. Only two types of quark – those labelled "up" and "down" – form the protons and neutrons of everyday matter, but physicists have to invoke three others to build up the more exotic particles created in high-energy collisions at accelerators, and when cosmic rays crash into the atmosphere. The table lists the six types of quark and some of their properties. To explain the behaviour of the more exotic particles, physicists have had to endow the quarks with properties not visible in the macroscopic world – strangeness, charm, and "bottomless" (sometimes called "beauty", but anyway "b" for short).

	Up u	Down d	Strange s	Charm c	Bottom b	Top t
Electric charge	$2/3$	$-1/3$	$-1/3$	$2/3$	$-1/3$	$2/3$
Strangeness	0	0	1	0	0	0
Charm	0	0	0	1	0	0
Bottomness	0	0	0	0	1	0
Topness	0	0	0	0	0	1
Baryon number	$1/3$	$1/3$	$1/3$	$1/3$	$1/3$	$1/3$

The quarks, like all other particles, have antimatter equivalents, in this case called anti-quarks, with equal mass but opposite properties such as electric charge. Some particles, including the proton and neutron, are built from a combination of three quarks, and are called baryons; others, collectively called mesons, comprise a quark–anti-quark pair, not necessarily of the same type.

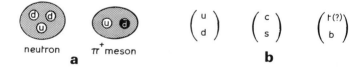

neutron π^+ meson

a

$$\begin{pmatrix} u \\ d \end{pmatrix} \quad \begin{pmatrix} c \\ s \end{pmatrix} \quad \begin{pmatrix} t(?) \\ b \end{pmatrix}$$

b

Three quarks make up a hadron, such as the neutron, while a meson contains a quark and an anti-quark. Their properties (or more strictly, their quantum numbers, see the Table) add up to give those we observed for the particles the quarks form.

"down" quark (charge $-\frac{1}{3}e$); the neutron, with charge zero, has one "up" quark and two "down" quarks. This fractional charge would be a clear indicator of a free quark. But this is not the whole story, since quarks possess properties other than electric charge (see Box 14.1). When Murray Gell-Mann and George Zweig proposed the existence of quarks they knew that there were "strange" particles that did not decay into other less heavy ones as readily as they should according to established theory. This had long been interpreted as evidence for the existence of a new property called "strangeness" – the name chosen reflected the ignorance of the processes involved. Just as electrical charge is found to be conserved in all known processes, so strangeness is found to be conserved in strong interaction processes. And it is this that restricts the way in which strange particles decay. They can do so only by the weak interaction and it takes much longer than it would by the strong interaction; the weak decays appear to be inhibited. So, the theory requires a third quark to endow the property of strangeness to the strange particles that it must inhabit.

In the 1970s, as Part Three describes, new varieties of particle were discovered, which required the existence of two more quarks, endowed with new properties, called "charm" and "bottom". With these five types, or "flavours", of quark, namely "up", "down", "strange", "charmed" and "bottom", we can build up patterns of hadrons using various combinations and predict the existence of new particles. Several such predictions have been verified in extremely difficult experiments at the most modern accelerators. Moreover, as Chapter 11 describes, it seems likely that there is a sixth type of quark, with a property called "top", and [in 1984] there is some evidence particles built from top quarks do exist.

But the behaviour of quarks is still more complicated than can be accounted for with five or even six "flavoured" quarks. Two related problems give rise to the need for yet more bizarre properties.

First, only certain combinations of quarks seem to occur in nature; three quarks for particles like the proton, and a quark–anti-quark pair for others like the pi-mesons. Particles corresponding to other combinations of quarks have never been found.

Secondly, for the combinations of quarks that apparently occur, one of the fundamental postulates of quantum theory requires that quarks of the same flavour, but in the same hadron, must be different, so the theorists assigned quarks an additional property, dubbed "colour". Each flavour of quark comes in three "primary" colours; red, blue and green. Observed hadrons can then be explained as combinations of quarks with no net colour; either they consist of a quark of each colour to form a "white" particle, or the colour of the quark is cancelled out by the "anti-colour" of its complementary anti-quark. The status of "colour" has progressed from theoretical speculation to become the key idea in the new theory of quarks.

The new theory of quantum chromodynamics (QCD), is based on the forces between the quarks that arise from their "colour", in the same way that the electromagnetic force between particles arises from the electric charge; hence the name "chromodynamics", in analogy to electrodynamics. It is important to stress that this colour has nothing to do with the usual meaning of the word; it is just a convenient way of labelling a property that has no counterpart in the everyday world. The force between coloured quarks is related to the exchange of "gluons", called so simply because they glue the quarks together. This mechanism is analogous to the interaction between electrons by the exchange of photons, described earlier. Because there are three different colours in contrast to just the one electronic charge, we need eight different gluons to account for the possible interactions. The gluons can themselves be coloured and this leads to interactions between them, too. This is in contrast to QED in which the photons are uncharged.

Physicists working at the Stanford Linear Accelerator Center in 1972 performed a series of now classic experiments that has reinforced our picture of quarks and the forces between them. In these "deep inelastic" experiments, beams of high energy electrons were aimed at proton targets. An electron emits a photon as it goes past the proton, and the photon then interacts with the proton. Although the scattering of electrons is observed in the experiments, the proton is being probed by the photons. By observing the scattered electrons at a given energy and at a certain angle, it is possible to "tune" the wavelength associated with the photon emitted by the electron.

When this wavelength is much larger than the proton, the proton appears as a point charge and the observed electron scattering is of a particularly simple kind. As the wavelength decreases to the dimensions of the proton the scattering becomes more complicated, as from an extended object. Then, when the wavelength of the photon is very small indeed, there is a magical change back to the simplicity of point-like scattering. The natural explanation of this simplicity, anticipated by the theorist James Bjorken of Stanford, is that the electron is now scattering from some point-line constituents within the proton, which could be the quarks.

To explain the experiments, the force between the quarks must be of a different nature from the other forces we know. The "deep inelastic" experiments show that the quarks inside the proton can be regarded as single, free particles, like marbles inside a bag. This means that the force between them must be very weak. On the other hand, no experiment has yet isolated a single quark; they are always bound together inside the hadrons, and this suggests that the force between them must also be very strong. To explain both, the force must change its character with distance. At the very short distances probed by the "deep inelastic" experiments, the quarks are effectively free from their forces, but when the quarks are separated by a relatively long distance, say the diameter of the proton, the force must have become very strong so as to confine them. So QCD has to explain two extremes of quark behaviour; freedom and confinement. The greatest success of QCD so far has been the successful description of the former; the greatest frustration, the unsolved puzzle of the latter.

The experiments have shown us that the strength of the colour charges must vary with the distance from which they are viewed. This idea is not new to physics, and a similar effect between electrical charges has been known about for a long time. When electrical charges are placed in water, each charge appears to the other to be diminished. This is because the electrical field surrounding each charge has polarised the water molecules, leading to a screening effect between them. When the charges are brought close together, so that there is no room for a screen, then the strength of each charge is effectively increased resulting in a greater force between them.

The same process can occur in empty space – the vacuum. Only naively does the vacuum live up to its name. In relativistic quantum theory it is a sea of virtual electron–positron pairs; the physical electronic charge we normally deal with is screened by the polarisation of this vacuum. When two charges are brought very close together, the effective value of the electronic charge increases.

A similar effect occurs in QCD, but with an important

difference. Now we are interested in the influence on the quark colour charges of the polarisation of the sea of quark–anti-quark pairs in the vacuum. As it happens, the presence of the gluon self-interactions produces an effect exactly opposite to the example from electrodynamics: the effective strength of the colour charges decreases as they are brought together and the quarks become "asymptotically free".

This striking behaviour was discovered by David Politzer of Harvard in 1973. Consequently Sidney Coleman and David Gross demonstrated that only gauge theories, like QCD, can exhibit this behaviour.

As if to show that there is really nothing new under the Sun, the unsolved problem of quark confinement also has a counterpart with which physicists are familiar, that is the problem of magnetic monopoles. Although there is no theoretical reason why we should not discover a single magnetic "north" pole, they maintain a mysteriously low profile. If we take a bar magnet and halve it successively, we shall never isolate the "north" end of the magnet, we shall just create smaller and smaller bar magnets with both "north" and "south" poles. The same is true of hadrons. No matter how hard we collide them together we never isolate a quark, but instead create new quark–anti-quark pairs.

The confinement problem has been studied in "thought laboratories" which theorists imagine to contain, for instance, only one space dimension and one time dimension, or in which the space-time continuum is divided into a lattice of discrete points (see Chapter 17). Results from these studies suggest mechanisms which might be at work, but conclusive results from the full theory have yet to be found.

One outstanding question, even while many of the fundamental problems of QCD remain unsolved, is what lies inside the quark? How are the quarks constructed? The "deep inelastic" experiments have been performed with photons with such short wavelengths that deviations from the simple, point-like scattering due to photon–quark collisions can be measured. These deviations are attributed to the emergence of a gluon from the quark just as the photon approaches, and this effect can be calculated within the framework of QCD. In other words, the closer we look, the more we see. These deviations represent a probing of the fine structure of the quarks themselves, but perhaps the effect is due to quarks and gluons and there is no need to introduce entities still more basic.

The hope of physicists studying QCD is that the mathematical structure of the theory will be rich enough to explain both the data from the scattering experiments and the fact that single quarks are never observed. QCD is a particularly favoured theory because it is very similar to the "field theories" of the other forces in nature, and this would be a great help in formulating "GUTs", the grand unified theories of the strong, weak and electromagnetic forces which currently challenge theorists.

1 March, 1979

15

Hybrids and glueballs

FRANK CLOSE

How the theory of the strong force allows for exotic new forms of matter.

Deep inside the atomic nucleus powerful forces are at work. The latest theories of these forces suggest that the energy locked up in the force fields may conglomerate and form exotic balls of pure energy, named "glueballs". Recent experiments may have already found the first examples of such glueballs, though the results could be interpreted in other ways. Theorists are now beginning to speculate that there might also exist further bizarre manifestations of glue-energy, in the form of particles dubbed "hybrids". But a hybrid of what? And what are glueballs, and how can you see them?

The possibility of such exotic forms of matter has only arisen as our understanding of the forces within the nucleus matured during the 1970s. In turn, these insights rest on the discoveries that the protons and neutrons within the atomic nucleus are built from quarks, and that the forces acting on quarks show similarities with electrical forces – though with subtle and far-reaching differences.

Atoms are held together by the electrical attraction between opposite charges: negatively charged electrons orbit a positively charged nucleus. The electric energy bound within the atom in this way can be transformed into electromagnetic radiation, such as visible light. Could there, therefore, exist "atoms" of pure light?

Nature does not allow this exciting possibility. Light is electrically neutral, so there is no attraction between opposite charges to cluster balls of light together and make exotic new matter. But the nuclear analogue of atoms made from light may well exist, in the form of glueballs. Indeed, the story of glueballs and hybrids provides a good example of how physicists often solve problems by drawing analogies with what they already know.

Quantum mechanics revealed one of nature's most exclusive secrets when in 1925 Wolfgang Pauli, then in Zurich, formulated his "exclusion principle", which states that no two electrons in an atom can share the same state of motion. This succinct rule so tightly limits the behaviour of electrons in atoms that only a few out of an otherwise infinite number of possible atomic varieties occur, and these turn out to be the very elements we find in nature! Thus with the advent of quantum mechanics the whole of chemistry had suddenly been subsumed in the new atomic physics.

By 1930 the quantum nature of the atom was common knowledge and attention turned to the nucleus, where a paradox emerged. The electrical attraction of opposite charges adequately explains the simplest atom, hydrogen, whose nucleus consists of a single proton which is orbited by a single electron. But in complex atoms as many as 92 protons are crammed together in a dense pack, seemingly contradicting the adage "like charges repel". What force prevents the nucleus from flying apart?

The Japanese theorist Hideki Yukawa suggested that a powerful attractive force operates within the nucleus. Reasoning by analogy he argued that if the electromagnetic energy locked up in the atom could be liberated as electromagnetic radiation in a staccato burst of particles – "photons" – so should the energy locked up in the nuclear force field be liberated as particles. These particles became known as "pions", and Yukawa's theory was vindicated when they were first detected in 1947 in experiments with cosmic rays.

But following hard on the success of this analogy were further problems. The smashing together of nuclei at high-energy particle accelerators revealed not only pions but scores of similar particles that no theory had predicted. By the 1960s the situation had reached chaotic proportions as experimenters had identified more than a hundred of these so-called "elementary" particles.

During the past two decades order has been restored to chaos and, yet again, analogies have been major guides. As atoms consist of electrons and nuclei, and as nuclei contain protons and neutrons (uncharged particles of about the same mass as protons) so are protons, neutrons, pions and their hundred siblings all built from more fundamental objects. Murray Gell-Mann, a theorist at the California Institute of Technology, named these more elementary particles "quarks" in 1964. In this picture, the proton and neutron are composed of three quarks, while Yukawa's pion contains only two – a quark bound to an anti-matter quark, or anti-quark. Because matter and anti-matter annihilate on contact the pion does not live long, unlike the proton and neutron which are stable in the nucleus over a long timescale.

The rule seems to be that quarks are bound together either in trios or as quark and anti-quark. This rule works so well that it must provide an important clue to the nature of the forces acting on quarks. Why are there no clusters of two, or four quarks? Why are no single quarks seen? Now I must confess at once that no one has yet given a complete theoretical explanation as to why isolated quarks are so improbable, or maybe even forbidden to exist, but if their imprisonment in certain clusters is a law of nature we can at least explain why only these particular groupings occur.

The first clue to the nature of the quark forces emerged soon after the idea of quarks was proposed, though it is only with hindsight that we can recognise it as such. Pauli's exclusion principle, which had so perfectly limited the electron states in atoms to give only the observed elements, should also apply to quarks. Yet Oscar (Wally) Greenberg at the University of Maryland discovered that it seemingly forbids the very clusterings that do occur. Given that in 1964 the quark idea was by no means universally accepted it is remarkable that Greenberg did not take the Pauli paradox as evidence that the quark model was wrong. Instead, he proposed an extremely radical resolution which, 20 years later, indeed appears to be correct. As this work is so pivotal I would like to spell out what the problem was.

Pauli's principle not only forbids two electrons to occupy the same state of motion in an atom, it also forbids two quarks to be in identical states in the proton or any of its siblings, such as the Ω^- particle. Now, the Ω^- (pronounced, omega-minus) consists of three quarks, each apparently identical to the other – a configuration that Pauli's principle excludes. To resolve the paradox Greenberg proposed that quarks possess a new sort of charge, which we now recognise to be similar to electrical charge except that it occurs in three distinct varieties. To distinguish between the different charges, physicists have whimsically referred to them as the red, yellow or blue variety and collectively they are known as "colour" charges. Moreover, instead of simply positive or negative values, as is the case for electric charge, there are "positive" and "negative" red, yellow and blue colours. Quarks carry positive colour charges, anti-quarks have the corresponding negative colour charges. (As "colour" is an arbitrary, but suggestive, name for this quark charge, the three colours could equally be red, blue, green, as other writers in this book have used.)

How does this solve the paradox of the three identical quarks in the Ω^-? The answer is that they need no longer be identical: if one quark carries red charge, one blue and one yellow they are quite distinct. Suddenly the Ω^- can exist − as experiments observe (Figure 15.1).

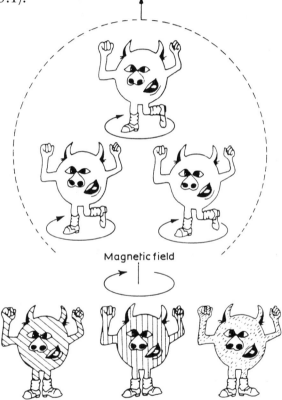

Figure 15.1 *The Ω^- particle should be built from three strange quarks, all spinning the same way, according to the basic quark model (top). but the Pauli exclusion principle forbids objects with spin $\frac{1}{2}$, such as quarks, to be in identical states at the same time. The resolution of the paradox is to endow the quarks with a new property, colour, such that the quarks come in three different colours (bottom). Then the three quarks in the Ω^- can be different although they spin the same way. (Credit: Frank Close)*

So in the mid-1960s we were asked to believe that quarks existed (even though no one had ever seen one) and, as if that wasn't enough to swallow, that they carried a bizarre new property, called colour. It is probably no surprise to learn that most particle physicists received these ideas with rather less than total excitement. But gradually evidence accumulated in favour of the colour hypothesis. In 1968 experimenters fired electrons into the heart of the proton and saw quarks indirectly. And by 1970 the first explicit manifestation of three-fold colour began to emerge. The ideas seemed to be right after all. Paradox had guided theory through analogy to coloured quarks, which did indeed seem to exist. In the next step, which brings us up to the present excitement over glue-matter, analogy seems to be providing a useful guide again.

Quarks, and the particles built from them, such as protons and pions, all experience the strong nuclear force. Electrons, which are *not* made of quarks, do *not* experience that force, which is why they are found on the periphery rather than in the nuclei of atoms. The coincidence is suggestive: could the colour charge of quarks be the source of the strong force? This would immediately explain why electrons are blind to the strong nuclear force and might explain the magic clusterings of quarks in trios or as quark–anti-quark.

The analogy proves fruitful. When the familiar rules of "like charges repel" and "unlike charges attract" are applied to the three varieties of colour charge, the attractions lead directly to quarks that cluster in trios or with anti-quarks (see Box 15.1). The forces that quarks feel do indeed appear to be analogous to electrical forces. Moreover, electromagnetic radiation in the form of photons has a colour analogue – the radiation of "gluons".

There is one subtle but far-reaching difference between gluons and photons. Photons do not themselves carry electrical charges and so a whole host of them can travel across space without getting in each other's way. Gluons, on the other hand, carry colour charges and attract each other as they travel across space. Indeed the colour forces pulling the gluons together are the same as those that bond the quark–anti-quark couples or quark trios. Experiments have revealed hundreds of examples of quark clusters, so should not clusters of gluons – "glueballs" – also exist? And if they do exist, where are they to be found?

Protons and pions are made of quarks, so when we smash them together we tend to discover further clusters of quarks. Most

Box 15.1   How colour forces work

Suppose that colour charges behave in a similar manner to electrical charges. Quarks carry positive colour charges and anti-quarks have negative charges, and "opposite charges attract". So a quark with positive red charge will attract an anti-quark with negative red (see Figure). Similarly positive yellow (or blue) attracts negative yellow (or blue). Thus do quarks and anti-quarks attract one another. Such bound states are well known; they form the variety of matter known as mesons of which Yukawa's pion is the most familiar example.

In electrostatics, two positive charges are always "like charges" and repel; in colour two red quarks are always alike and repel. But what about a red quark and a blue quark? These are alike in that they are both quarks, with positive colours, but unlike in that the colours differ. It turns out that the rule for attraction is generalised slightly. Not only "opposites" attract; but "unlikes" also attract. Thus a red quark and a blue quark can attract each other, as can red and blue, or yellow and blue. A third quark can be attracted

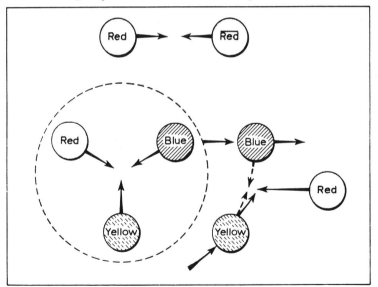

Opposite charges attract, so a "positive red" quark binds with a "negative red" anti-quark to form a meson (top). But "unlikes" attract" also, so quarks of three different colours can also form a particle (a hadron), which repels further quarks of the same colours.

Box 15.1 *continued*

only if its colour differs from each of the quarks already present. Thus a red–yellow–blue cluster can form (see Figure). Indeed, such a cluster *must* form, as the pairs cannot survive alone and always attract the missing colour to form a threesome.

These systems of three quarks exist quite happily, as protons and neutrons among other things. A fourth quark brought up to this trio is repelled by one of the three already present – the like colour – and so the forces are naturally neutralised once clusters of three have formed.

experiments in the 1950s and 1960s were of this kind, so physicists built up a whole menagerie of particles that we now recognise to be quark clusters. To find glueballs we must look for rather subtle effects, or perform different kinds of experiments.

In the 1970s physicists increasingly turned their attention to experiments involving electrons and anti-electrons (positrons) as projectiles. These particles are not built from quarks and, moreover, they can mutually annihilate upon impact. Out of the intense ball of energy thus created new varieties of matter might appear to which earlier experiments could have been insensitive. Such hopes have been fulfilled, exceeding the most optimistic expectations. Extremely massive particles, such as the psi (Ψ) and upsilon (Y) have been produced, which were previously unknown. These are not glueballs, but theorists suggest that up to 1 in every 10 psi particles will produce a glueball when it decays. Indeed some new particles seen in these decays are *prima facie* candidates for glueballs. These particles, named the iota (ι) and theta (θ), weigh in at about one and a half proton masses. They are born in the way expected for glueballs, and have the right sort of masses, but they die in ways that differ from those expected for glueballs. I wouldn't bet against one or even both of the particles being a glueball, but it is a brave person who would claim that glueballs have been found.

One point everyone seems to agree upon is that it may be very difficult to identify glueballs with certainty. How might we otherwise look for manifestations of gluons bound by colour forces? The new idea, which has begun to receive the attention of theorists on both sides of the Atlantic, is that colour forces could attract quarks and gluons mutually, forming hybrid objects.

The attraction of opposite charges pulls a red quark and a red anti-quark together to form a bound state where colours are neutralised – a familiar meson. But the rules for colour are more general: "unlikes attract", so a red quark can be attracted by a blue anti-quark. The positive red and negative blue do not mutually neutralise, but this pair can attract coloured gluons and so neutralise the colour forces. Thus a cluster of quark–anti-quark-gluon can form – in other words, a hybrid meson. These "half-quark" – "half-glueballs" were named "hermaphrodite" (as in biology) in Europe or "meikton" (from the Greek for a mixture) in the US. To gain uniformity the theorists have agreed to adopt the more staid name "hybrid" for this postulated class of matter. Now the aim is to find experimental evidence for it.

Unlike glueballs, which can be produced only in specific experimental conditions, hybrid particles could be produced in conventional particle-physics experiments. Already there are claims that such entities have been produced in past experiments, but not recognised. This is possible, but still far from certain. However, new problems may be confronting us.

The early studies of hybrids seem to be predicting too many states; the known particles fit the picture almost perfectly as quark clusters, there is no need for anything more. Some theorists are beginning to conjecture that hybrid states might not in fact exist even though naive theory suggests that they should. Are we thus *en route* to a new paradox? If so, are new profound insights soon to follow? Indeed there are deep subtleties in the way that colour forces behave that are only partially understood so far. The search for glue-matter is intensifying and its existence, or otherwise, seems likely to be a central question in the mid-1980s.

7 July 1983

16

The search for gluons

CHRISTINE SUTTON

Evidence for gluons, and for quantum chromodynamics, comes from a machine called Petra, an electron–positron collider at the German accelerator laboratory near Hamburg.

Nineteen seventy-nine was the year when the focal point of research in particle physics returned to Europe, after several years in the US, when fascinating results began to pour out of a laboratory in Germany. There, physicists from many countries, including the UK and the US, were finding evidence for *gluons* – the particles believed to "glue" together protons and neutrons inside the nuclei of atoms. These discoveries give impetus to the belief that physicists are at last on the right track with the theory known as quantum chromodynamics (QCD), evolved to describe the so-called strong nuclear interactions between neutrons, protons and other similar particles.

These results came from experiments at DESY, the German national high-energy physics laboratory on the outskirts of Hamburg, where PETRA, at present the world's largest working electron–positron storage ring is sited (see Box).

The most exciting results from PETRA emerged when the experimenters began to study hadrons – particles, such as protons and neutrons that interact via the strong force. The hadrons appear to be built from mere fundamental particles, the quarks. At PETRA, hadrons are created from the bursts of energy produced when beams of electrons and positrons collide and annihilate. The experiments were designed to discover how the emerging hadrons share out the energy of the annihilating electron and positron, and in which directions the particles go. For the previous five years experiments at smaller electron–positron machines had shown that hadrons produced in these reactions tend to be concentrated in

Box 16.1 PETRA – the gluon machine

PETRA stands for Positron Electron Tandem Ring Accelerator. The machine is based on a 2.3-km ring of pipe under extremely good vacuum, through which electrons and positrons (the positively-charged anti-matter counterparts of electrons) fly in opposite directions, eventually to collide at predetermined crossing points; magnets around the ring constrain the particles to their circular path. Before they collide the particles are "fed" with radio-frequency energy to accelerate them to an energy that can be up to 23 times the energy equivalent to the mass of a proton. When the electrons collide with the positrons, they *annihilate*, cancelling each other out to give for an instant a "puff" of pure energy equal to the *total* energy of the two beams. This energy can rematerialise as a variety of other particles, including electrons and positrons; it is carried by a *photon*, or "particle" of electromagnetic radiation.

The story of PETRA has been mainly one of success. Proposed in 1974, the machine's first electron–positron collisions took place in the autumn of 1978, six months ahead of schedule. Although dogged by difficulties that prevented them from achieving the rate of electron–positron collisions expected, DESY's technicians brought the machine's total energy to 36.6 GeV, close to its design energy of 19 + 19 GeV in the spring of 1980. (Prior to PETRA, the maximum energy reached in electron-positron collisions had been 5.1 + 5.1 GeV, at a smaller storage ring called DORIS, also at DESY.)

Experiments monitor four points around PETRA where the two beams of particles collide. At these crossing regions physicists wrap a variety of particle detectors around the vacuum pipe so as to see

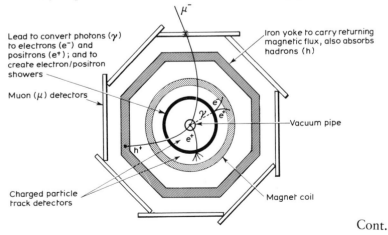

Lead to convert photons (γ) to electrons (e^-) and positrons (e^+); and to create electron/positron showers

Muon (μ) detectors

Charged particle track detectors

μ^-

Iron yoke to carry returning magnetic flux, also absorbs hadrons (h)

Vacuum pipe

Magnet coil

Cont.

Box 16.1 cont.

An experiment to study electron–positron collisions should contain detectors to identify different types of particles, others to track the "footprints" of the particles, and a magnetic field to bend the paths of oppositely charged particles in opposite directions. The picture shows the concentric arrangement of the JADE detector at DESY's larger storage ring, PETRA. (Credit: DESY)

what the annihilations produce. The basic idea is to use a solenoid coaxial with the beam pipe to provide a magnetic field to bend the paths of electrically charged particles leaving the collision region,

Box 16.1 cont.

and to have a detector that will record the tracks of these particles. Add some detectors to register neutral particles, and others designed to allow you to identify different types of particle and you are well on the way to a useful set of apparatus (see Figure). Four of the five experiments designed so far for operation at PETRA have been based on this formula, and complement each other by using different techniques.

two jets, spraying out from the collision point (Figure 16.1a). The quark model has a simple explanation for this. The electron and positron annihilate, producing a photon of the appropriate energy which then creates a quark, together with an anti-quark with exactly equal and opposite properties. The quark and anti-quark each materialise as hadrons which cluster about the direction of the parent quarks (Figure 16.1b). As the energy of the annihilation increases, the quark and anti-quark shoot off with more momentum, which is transmitted to the emerging particles. So the clusters of particles cling to the directions of the original quark and anti-quark, forming two well-defined "jets". The experimenters at PETRA clearly saw two jets when the machine's total energy was at 13 GeV and at 17 GeV.

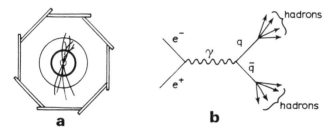

Figure 16.1 *The tracks of hadrons – particles made of quarks – produced in electron–positron collisions appear to form two jets (a), believed to come from a quark–anti-quark pair (b).*

However the nascent theory of the strong interaction – quantum chromodynamics, or QCD – suggests that this picture is not so simple at higher energies. According to QCD, quarks are bound together within a particle such as a proton by exchanging other particles called gluons – rather as players passing a ball are "held" together. The gluon carries the strong nuclear force between quarks. (This idea is analogous to QED's description of the interactions of electrons and other electrically charged, particles, which proceed via the exchange of photons.) QCD successfully explains why no free quarks have been found, as quarks are always held together by gluons. Looking for a free quark is like looking for a piece of string with one end.

One result of QED is that if you accelerate an electron it will emit photons, that is, electromagnetic radiation; this is how radiowaves are produced. By analogy QCD predicts that accelerating quarks should radiate gluons.

How does this affect the two-jet picture seen at PETRA's lower energies? A quark with barely sufficient energy to radiate a gluon will emit a gluon that is not itself very energetic. A gluon produced in this way can create a quark–anti-quark pair just as a photon does, and so ultimately the gluon will materialise as hadrons, as the quarks do. But if the gluon has little energy, the particles it creates cannot stray far and will intermingle with those of the parent quark; the effect is to broaden the jet of particles associated with a radiating quark (Figure 16.2a). With increasing electron–positron energy, more energy becomes available to the gluon and eventually the gluon can throw off particles quite separately from the quark (Figure 16.2b). Then three distinct jets should appear, corresponding to two quarks and a gluon.

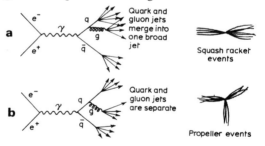

Figure 16.2 *(a) If a quark (q) radiates a low-energy gluon (g), the gluon's jet of hadrons will emerge with that of the parent quark, producing events that resemble a square racket. (b) With more energy the gluon's jet becomes distinct, giving rise to the so-called "propeller" events.*

What do the experimenters see? With PETRA running at a total energy of 30 GeV things had already changed; some 10 per cent of the events were not the same as the manifestly two-jet events that appear at lower energies. One of the jets in each of these rarer events is generally broader, and in some cases it is possible to see

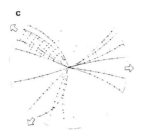

Figure 16.3 *A three-jet event recorded by the TASSO experiment.*

three clear jets. Figure 16.3 shows a three-jet event from the TASSO detector; physicists from Imperial College, London, Oxford University and the SERC's Rutherford Appleton Laboratory work with colleagues from Germany, the US and Israel, on this experiment that was the first to report the observation of events with three jets. Later all other groups at PETRA reported similar results.

It is important to realise that, at energies around 30 GeV, the three-jet structure is not usually clear to the naked eye. The experimenters have to devise careful tests to confirm that a model with three initial jets best describes their data – not two jets, or even four jets or more. Another important point is that events with three jets do not necessarily prove that one of the quarks produced in the electron–positron annihilation has radiated a gluon; as in most scientific "proofs" a strong chain of evidence has to lead to such a conclusion.

A first important clue comes from the fact that the particles in the rarer events tend to lie in the same plane – leading to events like "squash rackets" in the thin-jet/broad-jet case, and others resembling "propellers" when three jets are distinct (Figure 16.2). This planar structure is essential to conserve energy and momentum if the jets originated as three particles – quark, anti-quark and gluon.

Assuming that the unusual events seen at energies of 30 GeV or more are caused by quarks radiating gluons, the physicists at PETRA have calculated the strong coupling constant – this is the probability that a quark will radiate a gluon, and is a measure of the strength of the strong nuclear force. The TASSO group obtained a value of 0.17 ± 0.02 for the coupling constant at 30 GeV. Unlike the electromagnetic coupling constant – or the "fine structure constant" – which is constant at all energies with a value of 1/137.03, the strong coupling constant should, according to QCD, decrease with energy.

Whether the physicists at PETRA have indeed seen evidence of quarks radiating gluons, and whether QCD is indeed the correct theory of the strong interaction are questions still open to debate. Other evidence for the existence of gluons has come from lower energy data collected at the smaller electron–positron ring DORIS, also at the DESY laboratory. Around a total energy of 10 GeV physicists at DORIS found a particle called the *upsilon*, which lives for only an instant before decaying into other more stable particles. Through elaborate statistical analyses the physicists have been able to interpret the upsilon decays in terms of three jets of particles, in this instance believed to emanate from three gluons (Figure 16.4).

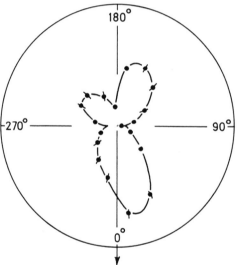

Figure 16.4 *Quantum chromodynamics (QCD) predicts that upsilon particles decay into three (or more) gluons, each of which then produces a jet of longer-lived particles. Physicists working on the Pluto detector at DESY's smaller electron–positron ring, DORIS, have analysed their data specifically to look for such decays. The team first finds the most energetic jet in an event and then makes a map of how much energy flows away in other directions. The three lobes on the plot show the upsilon is behaving as expected, just as QCD predicts.*

As for QCD, the least that can be said about the theory is that it provides the only common way of reproducing the results of all experiments. Many results can be explained without QCD – there are elaborate models that envisage the decay of the upsilon without needing gluons – but each explanation is different. The case for QCD and gluons, if not yet cast iron, looks very good.

11 September 1980

17

Computing the strong force

DAVID WALLACE

The unusual properties of the strong force make calculations of its effects difficult; but theorists find that techniques developed in other areas of physics are helping them to "tame" this force.

Of all the forces, the electromagnetic is the best understood. At the everyday level electric and magnetic phenomena are well described by the electromagnetic fields enshrined for more than 100 years in the equations of James Clark Maxwell's theory. But at the sub-atomic level, electromagnetism must be combined with relativity and quantum theory; a marriage from which the photon of light emerges as the quantum of energy of an electromagnetic wave. Moreover, one is led naturally to introduce another "matter" field whose energy quanta are the electrons themselves. The resulting theory of interactions between photons and electrons, developed in the 1940s, is called quantum electrodynamics or QED.

The key characteristic of QED is a property of the basic equations called "gauge symmetry". Most familiar symmetries reflect the fact that we can make changes to an object, such as rotating it through a certain angle, which do not change the object's appearance. Hence, we can rotate every point of a cube through 90° about one of its axes and it will appear unchanged. But if we rotate different points of the cube through different angles the shape becomes completely distorted. The defining feature of a gauge symmetry is that rotations, or more general symmetry transformations, *can* be different at different points in space (and at different times). So, because QED is a gauge theory we can make different transformations to the electron's field at different points in space-time without distorting our view of the physical world.

QED has proved extremely accurate, and its success has led to the idea that gauge theories might provide the correct description of all fundamental forces. This idea has come to maturity with the unified electroweak theory, for which Sheldon Glashow, Abdus Salam and Steven Weinberg received the Nobel prize in 1979. The electroweak theory is like QED, a gauge theory, although the symmetry of the theory remains hidden from view in

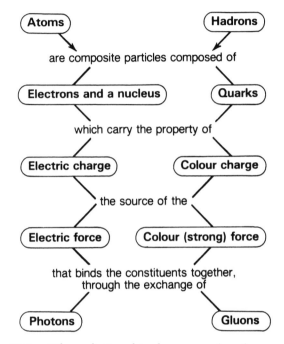

Figure 17.1   *The relationship between the electromagnetic force in atoms (left) and the strong force in hadrons (right).*

the everyday world, where the weak and electromagnetic forces appear manifestly different (Part Five).

It is now widely believed that the strong force is also described by a gauge theory and the specific theory, which many believe is the only credible candidate for the strong force, is called quantum chromodynamics or QCD. The theory deals with quarks (quanta of the "matter" fields, like electrons in QED) and gluons, which are quanta analogous to the photon. The gluons couple to quarks with a coupling whose strength is determined by an attribute of the quarks called "colour", hence "chromo"dynamics. This is just like the coupling of the electric field in the photon to the attribute of the electron that we call electric charge. The physical picture that lies behind QCD describes the proton as made up of three quarks, bound together by the forces generated by the gluons. At a superficial level, the idea is very similar to our understanding of atomic and molecular physics (see Figure 17.1). In atomic physics, neutral atoms are formed from a positively charged nucleus surrounded by a cloud of an appropriate number of negatively charged electrons. According to quantum theory, the electrons in a given atom can move only in certain specified states which have a discrete set of permitted energies. The atom itself therefore has a discrete set of allowed energies, the lowest of which is the stable atomic state, the higher ones corresponding to excited states of that atom. All the different states of all the different atoms can be obtained (in principle) by solving the quantum theory of the electric and magnetic forces between the electrons and the nucleus. The chemical forces that form molecules out of atoms are just a residue of the electric and magnetic forces responsible for forming the atom themselves. (These remarks in no way detract from the enormous practical difficulty of actually solving the quantum theory for complicated atoms or molecules.)

Similarly, we must now imagine that the proton is a state of three quarks bound together by a new kind of force. At the sub-nuclear level we are dealing with scales of distance, time and velocity such that it is essential to include relativity as well as quantum theory in our descriptions of the behaviour of quarks. Through Einstein's famous relation $E=mc^2$, the fact that the three quarks must be in allowed *energy* states (E) fixes the *mass* (m) of the proton and the hundreds of other "elementary" particles like it; they are just different combinations of quarks in different allowed states. In order to get the correct charges and other properties (strangeness, charm, spin, parity and so on) for these particles, there must be at least five different kinds of quark, each carrying half a unit of internal angular momentum (spin). We can then classify hadrons (the strongly interacting particles) as states of three quarks (baryons) or a quark–anti-quark pair (mesons). These states have zero net "colour", just as an atom has zero net charge. The nuclear force binding protons and neutrons within the nucleus should in this picture be viewed as the residue of the more fundamental strong force binding the quarks in the proton; nuclear physics is then like the "chemistry" of the strong force.

When one looks at the dynamical properties that this fundamental strong force between the quarks must have, it becomes clear that this analogy between the physics of hadrons and the physics of atoms is far from exact. This is highlighted by two paradoxical properties of the strong force between quarks. First, there is the problem known as "confinement". In atomic physics we can isolate a single electron from an atom, in a suitably energetic collision for example. However, with the exception of controversial experiments, it has proved impossible to knock an isolated quark out of any one of the known hadrons. The quarks appear to be permanently bound inside hadrons; increasing the energy of the colliding particles creates only quark–anti-quark pairs. These always reorganise themselves with the existing quarks so that only hadrons emerge from the collisions.

The second unusual property that a theory of quarks must confront is that of "asymptotic freedom". If we attempt to probe quarks at very short distances and times *within a hadron*, the quarks act during the short period of the collision as apparently *free* particles, as if unaware of the confining force. It is only over longer time scales (comparable with the 10^{-23} seconds that it takes to traverse the hadron at the speed of light!) that the strong confining forces come into play.

The gauge theory QCD has, almost uniquely, the necessary property for understanding this second property of the strong interactions. When quantum effects are included in the theory the effective coupling strength of the gluons to the quarks *decreases* with the distance at which we probe the quark. That this is, on the basis of our normal experience, a rather surprising property is worth stressing. It is useful to contrast it with the behaviour of a charged particle in a non-conducting medium, or dielectric. We know that a charged particle polarises a dielectric in such a way as to screen the electric field due to the charge. A positive charge, for example, becomes surrounded by a region of net negative charge which reduces the field from the original charge. The field determines the force on another point charge at some distance from the first, so the effect of the polarisation of the dielectric is to screen the first charge in a way that reduces the apparent coupling strength between the two charges. Of course, if the two charges approach one another to distances of less than an atomic spacing, no polarisation occurs and the coupling strength is no longer reduced. In this way screening effects in electromagnetism typically mean that the effective coupling strength *increases* at short distances. The opposite is the case in QCD.

The sceptical reader may query the validity of the contrast drawn between screening in a dielectric and asymptotic freedom in QCD. In a hadron, where are the analogues of atoms and molecules to provide the "anti-screening" (such that the coupling strength decreases with distance) to make the contrast meaningful? We can understand the answer qualitatively from the basic energy-matter equation, $E=mc^2$, of relativity, and from the uncertainty relation of quantum mechanics. This states that there is an uncertainty in the energy of a particle related to the time scale over which it is measured; the smaller the time scale, the greater the uncertainty. One can interpret these equations to mean that over sub-nuclear time scales (around 10^{-23} s or less) a ghastly mess of so-called "virtual" gluon and quark–anti-quark pairs may continually appear and disappear. These can be "colour" polarised by the strong force, just as the atomic charges of the dielectric are electrically polarised by the electric force, for colour is the source of the strong force just as charge gives rise to the electric force. It is the virtual gluons in this mechanism which, on "colour" polarisation by a quark, give rise to the *anti*-screening of the strong force.

I have discussed so far two properties of QCD that suggest that it might be the basis for the theory of strong interactions. It contains quarks that help to classify the hundreds of "elementary" particles; and the strength of the coupling of quarks in the theory becomes weaker as one probes their properties at short distances. The latter statement can be turned round: the effective coupling between quarks becomes stronger as one goes to larger distances.

Does the coupling become so strong in the theory that we can *never* pull an isolated quark out of a hadron, as most experiments would suggest? This question cannot yet be answered in the standard QCD theory because the only systematic calculational scheme available works well only if the coupling strength that characterises the force is a small number, as it is in QED where the technique is very successfully applied. Although the scheme can be used in QCD to establish consistently the asymptotic freedom of the effective coupling in collisions at short distances, it is useless in considering the problem of confinement, where the effective coupling is necessarily large.

In order to tackle confinement, we must reformulate QCD in a way that enables us to use new methods of calculation, and, in particular, schemes where the coupling can be strong. A key step in this direction was made by Kenneth Wilson from Cornell University in 1974. He showed how one could formulate QCD in

theoretical model in which the continuum of space and time is
replaced by a discrete four-dimensional lattice, most simply by a
"hypercubic" lattice – a four-dimensional lattice of equal
spacings. The "field" describing quarks becomes a variable at each
site of the lattice and the gluon "field" sits on each link joining
neighbouring sites. In the simplest version, the gluon link couples
with two quark variables at its end, and it enables a quark to hop
about on the lattice. This coupling also respects gauge invariance:
different transformations on the quark variables at each end of the
lattice are cancelled by tranformations on the gluon variable on
the link joining the quark variables. In other words, different
transformations can be made at different points in space–time
without distorting the overall picture, just as a gauge theory
demands.

Wilson's formulation of QCD as a "lattice gauge theory"
continues to be a source of great insight into the theory of strong
interactions. At the conceptual level it underlines the strong
resemblance between QCD and the conventional models of
statistical mechanics used to describe phase transitions. The latter
models are, of course, three-dimensional, and generally mimic
behaviour in crystalline solids, such as the atomic vibrations, or
magnetic or electrical properties. In these cases, the lattice really
exists. The introduction of a lattice into QCD, on the other hand,
is a mathematical artifact, but one that gives us new ways to
understand the theory and to make calculations. In the end,
provided we can make the lattice spacing small enough – that is,
much smaller than a hadron – we can expect to obtain a good
approximation to the original QCD theory based on a continuum
of space and time. A deep understanding of this intuitive
expectation is contained in the theory of phase transitions for
which Wilson was awarded the Nobel prize for physics in 1982.

Having formulated QCD on a lattice in a way that makes it look
like a problem in statistical mechanics (albeit four-dimensional),
we can immediately borrow successful techniques and apply them
to QCD. One such technique, which provides a good description
of behaviour in statistical mechanics at high temperatures, is called
the "high-temperature expansion". When it is applied to lattice
QCD it turns out to yield a calculational scheme valid at large
coupling strengths – the "strong-coupling expansion". Recall that
this is precisely the regime that we cannot describe using the
standard QCD theory.

Moreover, in the strong-coupling expansion a very simple

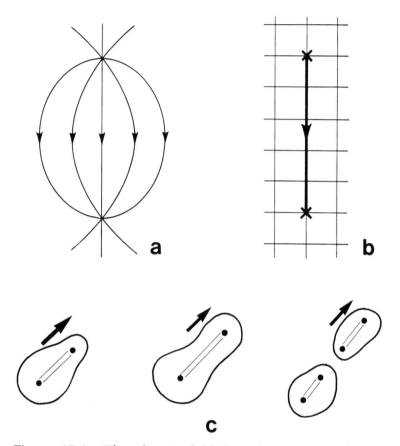

Figure 17.2 *The electric field lines between equal and opposite electric charges balloon outwards (a). But in the lattice model of the strong force the colour field between a quark–anti-quark pair at ultra-strong couplings emerges along links of the lattice (b). This picture gives us a way to represent the formation of hadrons in a collision (c), as a struck quark draws out behind it a tube of field lines, which splits when a quark–anti-quark pair is created.*

picture of confinement emerges. To highlight this, it is helpful to remember how the usual electric field lines look round a pair of equal and opposite charges (Figure 17.2a opposite). Figure 17.2b on the other hand, shows the *colour* field between a quark and an anti-quark in the lattice model at ultra-strong coupling. It does not balloon out like the electric field of 17.2c but essentially emerges from one quark along a series of the lattice links and ends on the anti-quark. Each link in this chain costs a certain energy. In this picture, therefore, the energy of the quarks and the gluon chain holding them together *increases* with the separation of the quarks. The quarks are permanently confined because it would take an infinite amount of energy to separate them to arbitrarily large distances.

This argument also gives a clear picture of what happens if we try to knock a quark out of a hadron, in a high-energy collision with another particle, say. The struck quark moves off rapidly, stretching out a gluon string behind it. At some point the energy stored in this string becomes so great that it pays to create a quark–anti-quark pair somewhere in the string: an extra hadron has been created in the collision, but no free quarks (Figure 17.2c).

It might seem that, with the quantitative results obtained from the strong-coupling expansion, one has achieved the ultimate goal of calculating the strong effects that confine the quarks. Unfortunately this is not the case.

To appreciate this rather subtle point, we must understand what the coupling strength between quarks and gluons in the lattice model actually means. In the sense discussed earlier in the context of asymptotic freedom, *it is the effective strength between quarks that are separated by a lattice spacing*. This becomes absolutely crucial when we remember that the lattice spacing ought to be small in comparison with the size of a hadron in order to obtain a good approximation to the original continuum QCD theory. We then realise that, if we wish to have a lattice spacing that is say 1/10 the size of a hadron (about 10^{-16} metres), we must adjust the coupling parameter in the lattice model to a *small* value, for in QCD the coupling becomes *smaller* at smaller distances. Only when the coupling in the lattice model is small is the hadron big enough to contain enough lattice points to ensure a good approximation to a continuum.

The results from the strong-coupling expansion are therefore inadequate. They imply that there is confinement in the physically artificial conditions where space-time is grainy because the lattice spacing is large. They do not guarantee that confinement takes place in the regime of physical interest where the spacing is small. Indeed, they had better *not* do so in general for the following reasons.

We know that we can construct a lattice model for the quantum field theory of the electromagnetic force, QED. This is similar to the model for QCD and likewise has a strong-coupling regime in which electrons and their anti-particles, positrons, would always be bound in neutral combinations. Fortunately, the lattice model for QED also has a regime in which electrons and positrons can emerge as isolated particles, as we observe. What happens is that the lattice QED model undergoes a phase transition, just as statistical mechanics describes for the phase transition from gas to liquid. At (unobserved) strong couplings the electric flux is arranged in tubes that ensure confinement; at weak couplings the electric flux has the familiar dipole form of Figure 17.2a – this is the "phase" of QED that is realised in the physical world.

Returning to quarks, gluons and QCD, does the confining phase at strong couplings persist for arbitrarily small values of the coupling? A rigorous proof has not yet been given but will probably be constructed following the recent work of E. T. Tomboulis, at Princeton University, on a simpler model with gluon variables only. Even after such a proof is given, however, there still remains the problem of actually calculating the masses and other properties of the hadrons.

Following the lead of Michael Creutz at the Brookhaven National Laboratory and Kenneth Wilson, many theorists are involved in an enormous effort to try to resolve these questions by numerical simulation of the behaviour of quarks and gluons in lattice QCD. The basis of this approach is conceptually very simple. It exploits the so-called "Monte Carlo method" which has already been successfully applied in statistical mechanics. In this approach one starts by assigning a set of values for the quark and gluon variables on each of the sites and links, respectively, of a four-dimensional lattice as large as computer resources permit. This "configuration" is then allowed to evolve in computer running time, part by trial and error, in order to obtain a sequence of configurations. If the lattice spacing is small enough in the sense discussed above, a typical configuration generated in this way should be a good approximation to the kind of mess of virtual quarks and gluons that is the true state of empty space – the vacuum – in the physical world. One then introduces some hadronic particle into such a configuration in the form of a *specific* combination of quarks and gluons, and measures the probability that it moves a certain number of steps along some lattice direction, propagating through space-time. Theory predicts that the propagation for a particle of mass m will decrease in a way that depends on the product of m and the lattice spacing a. Thus for each particle one should be able to read off the quantity ma from a "propagation function" measured on the lattice. In this way the model provides a key to the masses of the observed particles.

Many other physically interesting properties can also be calculated in this way. For example, one can calculate how the energy of a quark–anti-quark pair depends on their separation. According to the "confining tube" picture I discussed earlier this energy should increase linearly with the separation, for large separations. One may therefore try to measure the string "tension"; that this should always be non-zero is one of the tests for confinement.

The computing resources necessary to produce reliable results are, with techniques presently available, enormous. First, we must be able to store and work with the 3×3 matrices of complex numbers that must be used to represent the gluon variables; there are four of these for each site of the lattice (one for each of the links in the four directions). Secondly, we wish to make the *linear* dimension of the lattice stored in the computer as large as possible, so as to maximise the number of points. But the computing time for a single step of the whole lattice increases with volume – double the size of the lattice and the time-step takes 16 times longer. Thirdly, we are simulating the coherent behaviour of structures (the hadrons) extending over many lattice spacings and this demands running the program for a very large number of time-steps. Fourthly, the discussion on quark variables above glosses over the fact that they behave in a way that means they cannot be represented by a simple number. Methods for overcoming this problem have been proposed and are currently being tested; but they may demand an increase in computing time.

Results to date are encouraging. A numerical solution of the strong-interaction problem seems feasible, but we are still some way from obtaining reliable data in which systematic effects are fully understood and controlled. Figure 17.3 p. 174 gives some idea of the quality of results so far.

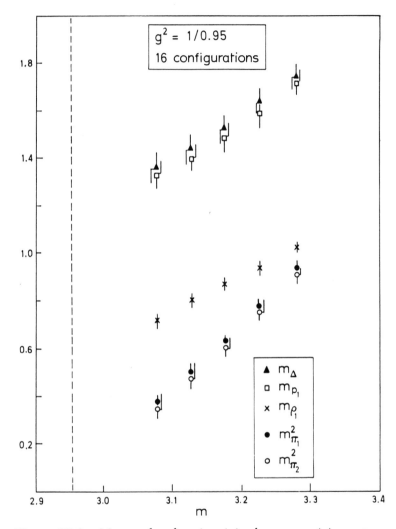

Figure 17.3 *Masses for the pion (π), rho meson (ϱ), proton (p) and delta baryon (Δ) as a function of the parameter* m *that specifies the quark mass in the lattice calculations. (All masses are in dimensions of lattice units.) The dashed line pinpoints the physical value of* m *derived from the observed small value for the mass of the pion.*

Figure 17.3 shows results for the masses of some hadrons as a function of the parameter *m* which controls the quark mass, at a fixed value of the coupling strength. These data are from an analysis of 16 independent configurations on an 8 × 8 × 8 × 16 lattice, using roughly 100 hours of running time on the Distributed Array Processor at Edinburgh University. The results illustrate the problems faced in this kind of calculation.

The lowest set of points are for a combination of quark and anti-quark, the pi-meson, or pion; all data must be extrapolated to that physical value of the quark parameter, *m*, at which the pion mass is almost zero, in line with its experimental value. From the extrapolated value of the data for another meson, known as the rho, one can estimate, using the known physical value of the rho's mass, the lattice spacing in units of 10^{-15} m (or fermis). The data in Figure 17.3 suggest that for this value of the coupling strength, the lattice spacing is 0.133 ± 0.010 fermi. Our box of eight lattice units in spatial dimensions is then roughly one fermi – or just about large enough to contain a pi- or rho-meson. The problem is that the masses for the proton and delta baryons are too close to one another and too heavy compared with the rho meson. Such features are typical of the strong-coupling regime.

To improve these results one has to decrease the coupling strength in order to see the continuum behaviour. However, one would then need to enlarge the size of the lattice if it is to contain a hadron. An alternative approach, which we at Edinburgh and other groups are studying, is how to modify the basic lattice model to make it intrinsically closer to the continuum theory.

The optimism generated by earlier results for hadron masses on smaller lattices has now been tempered by the recognition that reliable calculations require substantially greater computation. One can be sure that the theoretical ingenuity of physicists will be matched in this work by practical enterprise to acquire ever more powerful computing resources.

1 December 1983

PART FIVE

A step towards unity

The structure of matter is intimately linked with the nature of the forces that bind it together. We have seen in Part Four how an understanding of the quarks that form the nuclei of atoms depends in turn on a comprehension of the strong nuclear force that keeps them there. The same kind of relationship holds even on cosmic scales, where an understanding of how stars and galaxies form stems from a knowledge of the gravitational force that pulls matter together across the vastness of space. We cannot, on any scale, hope to appreciate fully the structure of matter without also learning about how forces operate.

Physicists identify four fundamental forces. Two of these – gravity and electromagnetism – have been studied in one way or another since antiquity. The weak and strong nuclear forces, on the other hand, have come to light only with studies of sub-atomic phenomena, particularly over the past 50 years. Part Four has brought us up to date with theoretical ideas on the workings of the strong force, but historically this was the last of the four forces to succumb to a reasonable theory.

Gravity came first under theoretical scrutiny with the renowned work of Isaac Newton in the 17th century. Later, in the first decades of the 20th century, Albert Einstein developed a more fundamental insight with his theory of general relativity, which reveals gravity as a warping of space–time, but that is a subject beyond the scope of this book. As for electromagnetism, that was first brought to order in the 19th century, with the work of James Clerk Maxwell. Then, in the 1940s, Maxwell's electromagnetic theory was adapted to deal with the unusual quantum world of the atom and sub-atomic particles. Thus was born quantum electro-dynamics, or QED as Paul Davies describes in Chapter 18. This is the quantum theory of electromagnetic phenomena and it is the most successful theory physicists have at their disposal. In the same chapter, Roger Blin-Stoyle records the award of the Nobel prize to three of the men who helped to make QED such a good theory

The success of QED led theorists to use it as a model in their attempts to come to terms with the other forces; in particular, to begin with, the weak nuclear force. This is the force that controls the nuclear reactions that fuel the Sun; it is also the force that allows the disintegration of unstable particles, and therefore underlies many aspects of radioactivity. In Chapter 19 Norman Dombey outlines the way that theorists were led by similarities between weak and electromagnetic interactions to develop a theory that incorporates both these kinds of outwardly disparate phenomena. The scheme thus echoes Maxwell's work of a century earlier, in which he had unified electromagnetism with light. As Paul Davies explains in more detail in Chapter 20, the new "electroweak" theory is analogous to QED. Indeed, it predicts the existence of "heavy photons" – massive particles that must mediate the weak interactions just as the photon of light in a sense carries the electromagnetic force between interacting charged particles.

According to the electroweak theory, one of the new mediating particles must be neutral and will give rise to a class of weak interactions known as "neutral current" interactions. When the theory was first put forward in the early 1970s, such interactions had never been seen. Their discovery in 1973, in the reactions of neutrinos in a hugh bubble chamber, was the first concrete evidence for the electroweak theory, as Fred Bullock describes in Chapter 21. But it was 10 more years before an accelerator could provide collisions of sufficient energy to release the carriers of the weak force – the charged W particles and the neutral Z particle. The discovery of these particles, formed in the fusion of matter with anti-matter, is described in Chapter 22.

The observation of neutral currents and the W and Z particles provides strong support for the electroweak theory. But it is not quite the end of the story. The weak and electromagnetic forces appear so different because the W and Z particles are heavy (80 to 90 times the proton's mass) while the photon has no mass at all. A full symmetry may exist between the forces only under very hot and energetic conditions, for instance, early in the formation of the Universe, where the mass of the W and Z would be negligible. But in our modern low-temperature Universe the two forces are manifestly different: the symmetry is "broken", or hidden from view. Andrew Watson describes in Chapter 23 how theorists have overcome the problems raised by this "symmetry breaking" in electroweak theory, and shows that another particle, called the Higgs after its inventor, awaits discovery if the theory is indeed correct. Discovery of the Higgs is crucial to electroweak theory; it is equally important for newer theories that seek to unite more forces, as Part Six will show.

18

QED: the electromagnetic force between particles

PAUL DAVIES

The theory of quantum electrodynamics is remarkably accurate and it ranks as the best theory that physicists have. But its development was not without problems, some of infinite proportions!

Ask any physicist what is the most successful physical theory and the chances are he will say quantum electrodynamics (QED for short). There is no doubt that QED represents the mathematical description of nature at its very best. I would like to explain what QED is, the problems it has encountered and surmounted, why it works so well, and how it is serving as a model from which to build a grand unified description of the forces of nature.

Electrodynamics deals with the motion and interaction of electrically charged matter, and the behaviour of the electromagnetic fields that these motions create, especially electromagnetic waves such as light and radio waves. Electrodynamics embodies both the laws of electromagnetism and mechanics and, in modern form, it also incorporates the theory of relativity (indeed, it was the midwife of relativity). Based on the celebrated equations of James Clerk Maxwell, electrodynamics provides an elegant and powerful mathematical framework for the description of a wide range of familiar phenomena, from the operation of a dynamo to the generation of radio signals. But on a subatomic scale the theory fails and Maxwell's electrodynamics has to be modified. Light plays a central part in both quantum theory and relativity, and it was only with the mathematical fusion of the two, into quantum electrodynamics, that detailed calculations of subatomic processes involving charged particles such as electrons and protons could be performed.

Quantum theory began in 1900 with Max Planck's brilliant idea

Plate 18.1 *When James Clerk Maxwell wrote down his theory of electromagnetism in 1862, it not only dealt with the connections between electrical and magnetic phenomena that experimenters such as Faraday and Oersted had already observed, but it also encompassed light, the visible vibration of the electromagnetic field. (Credit: Vivien Fifield)*

that light, and all electromagnetic energy, can only be emitted and absorbed by matter in discrete lumps or *quanta*, called *photons*, each of which carries an energy $h\nu$, where h is Planck's constant and ν is the frequency of light waves. It was, however, three decades before physicists arrived at any precise concept of exactly what a photon is. Light is the visible vibration of the electromagnetic field. The motion of fields differs in a fundamental way from that of a collection of particles because the field, being continuous, can move in an infinite variety of ways; it has "an infinity of degrees of freedom".

Sound waves share this property, and the infinite variety of motions of, say, the air are experienced as the infinite variety of sounds. Any particular sound can be analysed into a collection of pure notes; changing the sound quality reflects a different admixture of notes. Each note is called a mode. In the same way, light can be decomposed into a collection of modes; light of any colour is ultimately composed of a spectrum of primary colours or modes. The idea may be extended to all electromagnetic vibrations, even outside the visible region.

Vibrations of the electromagnetic field are rather like the vibrations of a mechanical oscillator such as a pendulum, although the latter is confined to a particular place while the electromagnetic field is spread throughout space. But mathematically the two types of vibration are identical, so the behaviour of the field may be studied by first considering the behaviour of mechanical oscillators.

It might be supposed that the admixture of colours to form multichrome light could be made in an unrestricted way, but in the 1920s scientists realised that an atomic-sized oscillator does not behave merely like a scaled-down version of a macroscopic oscillator, such as a pendulum. In particular, the energy of vibration of an "atomic" oscillator is quantised: it can possess only certain fixed multiples of a fundamental unit of energy, $h\nu$, where once again ν is the oscillation frequency. Similarly, each mode (frequency or "colour") of the electromagnetic field can have only fixed multiples of the energy $h\nu$. Each integer multiple is identified with a photon, so the photon is regarded as a quantised excitation of a field mode, and we see that Planck's original discovery of the discrete photon is simply an example of the quantisation of vibration.

Despite this mathematical success, regarding a photon as a

"field mode excitation" does not mesh easily with the intuitive image of a little "glowing ball". What endows photons with a particle aspect is the way in which they interact with matter. When the field vibration, which is extended through space, delivers its energy to an atom, it behaves very much as though the energy is concentrated in a localised region, just as in a particle.

The quantisation of the electromagnetic field – by which we regard it as an infinite collection of colours or modes, each of which has certain fixed lumps of energy – leads immediately to a problem when the *ground state*, or state of minimum field energy, is considered. Returning to the pendulum analogy, if we want to remove as much energy as possible we just need to hold the bob still at the middle of the arc of swing: in this way all the vibration energy is apparently removed. However, when we consider atomic-sized oscillators, quantum effects again become important. One fundamental quantum rule is Werner Heisenberg's *uncertainty principle*, which states that a particle cannot simultaneously have a well-defined position and a well-defined motion. If one quantity is fixed, the other acquires a random and uncontrollable value. This causes a problem if we want to sap all the energy from an "atomic" pendulum, because if we restrict its location by holding it in the middle, it will acquire random kinetic energy, and if we try to keep it still it will shy away from the vertical and acquire potential energy. Either way there will be a residual "zero-point" energy.

An analogous situation occurs with all the field mode oscillators of an electromagnetic field: each contains, even in its state of minimum energy (or ground state) a definite amount of zero-point energy. In the old classical theory, if all the energy is removed from a field (for example, if a light is switched off) then the field vibration ceases. But because of the Heisenberg principle, quantum field theory requires even black, empty space to contain some electromagnetic field vibrations – the zero point oscillations of each field mode. The consequences of this requirement are profound because there are an infinite number of modes and an *infinite zero-point energy* present, even in a vacuum. What does this mean?

Within the theory of electrodynamics, energy as such cannot be measured; only changes in energy are observable. So, for a practical calculation one is always free to rescale, or *renormalise* the zero point of energy, even by an infinite amount, without affecting the values of the (measurable) energy differences. In this way the alarming infinite zero-point energy is swept under the carpet. So long as we remain within the confines of special relativity this manoeuvre is successful and the predictive power of the theory is untarnished.

Tougher infinity problems emerge when the electromagnetic field is allowed to interact with electrically charged particles, such as electrons. Actually, there is a classical precedent here connected with the puzzle about the internal structure of the electron. If the electron's charge is distributed over a finite volume, then each region of charge will feel the electric field of the other regions, and there is a mystery about how the electron holds itself together against this mutual electric repulsion. If on the other hand, all the charge is concentrated at a point, as experiment indicates is apparently the case, then the electron has no internal parts to co-exist in equilibrium. But then another problem arises. As the volume of the charge distribution shrinks to zero, the electric field at the surface, and hence the electric repulsion energy, rises without limit. So, if the charge were truly concentrated at a single point, its energy would be infinite. According to special relativity, this energy has mass, so, added to its ordinary or "bare" mass, an electron has an infinite "electric" mass. But we can never switch off the electric field to observe the bare mass directly, so in practice what is actually measured is the total mass, bare plus electric, and this is, of course, finite. In this way, the bare electron mass is "renormalised" to give the physical electron mass.

But in the quantum theory of the electromagnetic field there are still more complications. This is partly because the quantum version of an electric particle is very subtle. When it was discovered in the 1920s that electrons and other subatomic particles possess wave-like properties, the behaviour of photons and electrons seemed rather similar. The "matter waves" that are associated with particles are not waves of any substance but waves of probability: the strength of the wave at a point is a measure of the likelihood of finding a particle there. Despite this physical difference, a matter wave and an electromagnetic wave are closely related mathematically, especially if the material particles are "relativistic" (moving with a velocity close to that of light). There does, however, seem to be one vital difference. Albert Einstein's famous equation $E = mc^2$, which predicts that energy possesses mass, also permits energy to be used for the creation of new matter. One example occurs when an energetic gamma ray (that is, a very short wavelength photon) suddenly disappears to be replaced by a pair of material particles, such as an electron and its anti-matter counterpart, a positron. Thus, neither photons nor electric particles are permanent entities, but may come and go in various numbers. How can we describe mathematically this creation and disappearance?

Electromagnetic radiation can be produced by violently disturbing electric charges, a process described by quantum electrodynamics as the appearance of new photons, resulting from excitations stimulated in the electromagnetic field modes by the rearrangement of the charges. If this analogy is followed for the matter field, it suggests that material particles such as electrons should also be regarded as quanta, that is as excitations of modes of vibration in the "matter fields". Because the existence of matter waves is already a quantum phenomenon, this further step is sometimes called *second quantisation*. The reward is a mathematical model in which excitations from one type of field can pass to another in much the same way as sympathetic vibration can transfer energy from one guitar string to another at resonance. Thus, the production of an electron–positron pair by a photon is regarded as the delivery of a quantum of energy into a mode of the electron–positron field from a mode of the electromagnetic field.

With this mathematical machinery, physicists can compute a vast array of electrodynamic phenomena. The coupling which facilitates the energy transfer is the electric charge, e, carried by the subatomic particle. All elementary particles carry the same numerical value of charge, so e acts as a fundamental natural unit of electricity. The strength of the coupling is the quantity $2\pi e^2/hc$ (c is the speed of light) which is a "pure" number, that is dimensionless, close to $1/137$. The smallness of this number implies that the exchange of photons between electric charges and the electromagnetic field produces only a minor disturbance on their motions, so it may be treated mathematically using the theory of small perturbations.

What this means is that as a first (or simplest) approximation we may compute, say, the emission of a photon by an atom, by ignoring the effect of that emission back on the atom. In the next "order" of approximation (about $1/137$ times smaller in effect) the back reaction is incorporated, but its additional effect is ignored. Continuing this way, a mathematical series of successively diminishing terms can be developed. The success of quantum electrodynamics rests crucially on the existence of this *perturbation series*, and the happy circumstance that e^2 is such a small number. In the 1940s Richard Feynman invented an elegant diagrammatic representation for the terms of this series. For example, Figure 18.1(1) shows the "first approximation Feynman diagram" for the emission of light by an accelerating charge; Figure 18.1(2) involves a higher approximation. Loosely, the first corresponds to the emission of one photon and the latter to two-photon emission.

Despite dazzling successes in computing the probabilities of processes like these (and many more), some diagrams produce absurdities because they give rise to infinite quantities. Consider, for example, Figure 18.1(4). This is a "second approximation process" (one emission and one absorption) in which a photon is created by one electrically charged particle and absorbed by another. Physically, the diagram describes scattering of the two charged particles off each other (like a collision between billiard balls), so we may regard the intermediate photon as the carrier or messenger of the electromagnetic field surrounding the charges. In Figure 18.1(5) a related "second order" process is shown, this time where an electron emits and reabsorbs the same photon. Processes of this type are occurring all the time, even around a stationary charged particle, because the Heisenberg principle, which states that momentum can be uncertain in value, also allows energy to fluctuate for short durations, enabling a photon to come fleetingly into existence so long as it is quickly reabsorbed. Thus, all charged

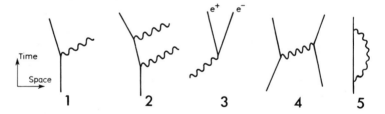

Figure 18.1 *Richard Feynman invented these diagrams to represent the interactions between particles and fields. Each element corresponds to a specific contribution to a mathematical description of the process that the diagram depicts. Solid lines represent particles, wavy lines the carriers of the force between them, and junctions depict the interactions. Thus diagram (1) may represent the emission of a photon by an electron) (2) the emission of two photons; (3) the formation of an electron–positron pair from a photon between two electrons; and (5) the emission and reabsorption of a photon.*

particles are dressed in a cloud of these so-called "virtual" photons, rather like ephemeral bees round a subatomic hive. The energy flowing around the loop shown in Figure 18.1(5), due to this restless activity may be readily computed. As the process is second order – in other words involves two interactions and terms of e – it should give a small correction; instead it is infinite! The trouble, once again, concerns the infinite number of electromagnetic field modes, with the virtual photons piling up ever greater quantities of interaction energy. The bare particle at the centre of the photon cloud is never observed in isolation, so the measured mass is, once again, the bare mass plus an (infinite) electromagnetic mass, the latter due this time to the cloud of virtual photons. So the infinite electromagnetic energy (or mass) apparently possessed by an electron is absorbed into the total mass and seems to have no physical consequences. Other diagrams give rise to infinities that can also be removed by "renormalising" other physical quantities.

Successive approximations, that is more complicated Feynman diagrams, throw up more and more infinities, but the striking feature of quantum electrodynamics (QED) is that all members of this interminable sequence can be absorbed into a small number of physical constants, such as the electron mass. This happy state of affairs is expressed by saying that the theory is "renormalisable" – in mathematical terms, the meaningless infinities do not in practice pollute the computation of observable quantities. And once physicists learned the correct rules for renormalising QED, they could compute small and subtle effects.

26 July 1979

Three physicists and the troublesome infinities

The Royal Swedish Academy awarded the 1965 Nobel Prize for Physics jointly to Professor Richard Feynman (California Institute of Technology), Professor Julian Schwinger (Harvard University) and Professor Sin-itiro Tomonaga (Tokyo Kyoiku University) for the major part that they played in the development of quantum electrodynamics.

It is now nearly 65 years since Planck first postulated the idea that an oscillating system is quantised; that is to say, it can only have certain allowed energies which are determined by the frequency of the oscillation and the quantum constant *h*. He was led to this idea in an attempt to explain the spectrum of electromagnetic radiation emitted by a hot body. The postulate was the first bold step in the formulation of a quantum theory of matter and radiation.

A major step forward in quantum theory was made in 1927 by Paul Dirac, who set up a theory of the emission and absorption of electromagnetic radiation by charged particles. This theory marked the beginning of quantum electrodynamics proper; in it, not only were the particle motions quantised but also the electromagnetic field itself. At about the same time, Dirac also formulated a theory for the electron which took account of relativistic effects. The theory was immensely successful and, among other things, was able to account for the magnetic properties of the electron and the fine details of the hydrogen atom spectrum.

Alongside the tremendous successes of quantum theory, there were profound conceptual difficulties. Werner Heisenberg's Uncertainty Principle allows the possibility that energy is not conserved in a system over short time periods; as a result an electron can emit transitory (virtual) photons which contribute to its energy and therefore to its mass. Unfortunately it turns out that the contribution to the energy is infinite (see last article by Paul Davies). However, theorists realised that, since these unacceptable excursions to infinity were always present, it was necessary to redefine the mass and charge in the theory so that from the outset only the *measured* electron mass and charge appeared. This process is known as mass and charge *renormalisation*, and serves to remove the unwanted infinities. The procedure was first carried out in a way consistent with the theory of relativity by Schwinger in 1947, and his calculations showed that, on taking effects of this kind into account, the magnetic moment of an electron should be slightly, but significantly greater than that given by Dirac's theory. Schwinger's calculations also showed that the fine structure in the hydrogen spectrum should differ slightly from Dirac's predictions. His results were found to be in excellent agreement with measurements made at that time by Foley and Kusch and by Lamb and Retherford. The calculations were not, however, completely consistent with the requirements of relativity theory, and further independent work by Tomonaga and Schwinger was needed before a theory completely satisfactory in this respect emerged.

Subsequent to this work, Feynman brought a new viewpoint to bear on the problem. The philosophy behind the approach of Schwinger and Tomonaga was that the interaction between charged particles was transmitted through the electromagnetic field by local action – the particles and the field quanta (photons) were treated on an equal footing. Feynman's approach on the other hand was to set up a description in terms of an "action at a distance" interaction between particles, in which the fundamental viewpoint and language was that of scattering. In particular, the formal use of techniques involving "Feynman" diagrams, each corresponding to a different possible virtual process, considerably simplified the executions of actual calculation. It was left to Freeman Dyson in 1949 to show that the theories of Schwinger, Tomonaga and Feynman were completely equivalent.

The success of quantum electrodynamics in accounting for the results of experiment is remarkable, and agreement between theory and experiment to about 1 part in 100 million has been achieved in the case of the interaction between electrons and the electromagnetic field. For this reason alone, it may be assumed that the theory will last. Nevertheless, the presence of the infinite quantities which are formally removed by the renormalisation procedure is worrying.

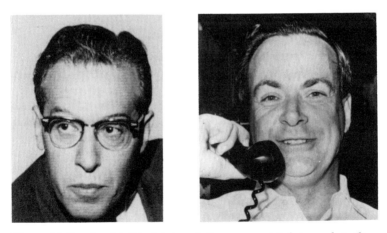

Plate 18.2   *In 1965, Richard Feynman (right) and Juilian Schwinger shared the Nobel prize for physics together with Sin-itiro Tomonaga, for their work on quantum electrodynamics, the quantum theory of electromagnetic phenomena.*
(Credit: AP)

The development of basic quantum theory and in particular of quantum electromagnetic phenomena has led to the award of many Nobel prizes and it is fitting – many would say, high time – that the contribution of Feynman, Schwinger and Tomonaga to the theory of quantum electrodynamics should be recognised. What of the men themselves? Feynman is flamboyant, restless, energetic – a great expositor and a star performer at any conference. Many true and apocryphal stories circulate about him, ranging from illustrations of his expertise as a safe-breaker to his virtuosity on the bongo drums! Schwinger, by contrast, is shy, quiet and precise; it is said that he always arranges to finish a lecture by the door so that he can escape from students. He is also reputed to have learned calculus as a very young boy from the *Encyclopaedia Britannica*. Of Tomonaga not so much is known. Apart from a time spent working with Heisenberg, he has spent little time away from Japan. Together they represent a remarkable trio, and the world of physics can only applaud the award of the Nobel prize to them.

ROGER BLIN-STOYLE
28 October 1965

19

Thoughts on unification

NORMAN DOMBEY

Attempts to unify apparently disparate phenomena can lead physicists to a better general understanding.

In the past, unification of phenomena that at first sight appear independent has signified great progress in physics. The classic example is the unification of electricity, magnetism and light. At the beginning of the 19th century, scientists had considered magnetism and electricity to be independent. But in 1819 Hans Christian Oersted showed that a steady electric current generated a magnetic field, and in 1831 Michael Faraday showed that a time-varying magnetic field would generate an electric current in a conductor. Together these results produce the combined subject of electromagnetism.

In 1862, James Clerk Maxwell wrote his famous paper in the *Philosophical Magazine* in which he assumed that a time-varying electric field (the displacement current) would also generate a magnetic field. This led to his prediction that electromagnetic waves existed and would propagate at a velocity c equal to the ratio of electromagnetic to electrostatic units of measurement. Numerically c turned out to be remarkably close to "the velocity of light in air, as determined by M. Fizeau" which was "70 843 leagues per second". Maxwell concluded that "we can scarcely avoid the inference that light consists of the transverse undulations of the same medium which is the cause of electric and magnetic phenomena". Thus the theory of light was unified with the theory of electromagnetism, although it took another 30 years before Heinrich Hertz was able to demonstrate positively that electromagnetic waves did exist.

In a somewhat similar vein nuclear physics contains examples of processes that appear to have little in common, but which may be unified. An atomic nucleus, for example, can in certain circumstances emit a gamma ray, or high-energy photon, in a process called gamma decay; this is an electro–magnetic effect. But a nucleus can in other circumstances emit an electron in a process known as beta decay, which occurs via the weak nuclear interaction (Figure 19.1a).

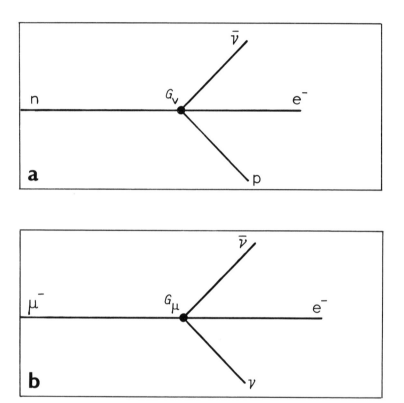

Figure 19.1 *Enrico Fermi thought of the beta decay of a neutron (a) as occurring at a point, with the coupling constant G_v giving a measure of the strength of the weak interaction. Muon decay (b) is a similar process with, it turns out, the same strength, so $G_\mu = G_v$.*

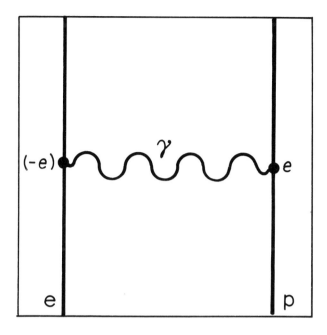

Figure 19.2 *Electromagnetic interactions occur via the exchange of a photon, γ.*

The two decays seem quite different. The strength of the electromagnetic interaction is much larger than that of the weak interaction, and the range of the electromagnetic force is much greater than that of the weak force. The probability of the beta decay of a neutron into a proton, an electron and an anti-neutrino is given by the so-called Fermi weak coupling constant, G_v. On the other hand, the electromagnetic interaction between an electron and a proton, which arises from the exchange of a photon between the particles (Figure 19.2), is given by the product of electric charges of the particles, that is magnitude e^2, where e is the unit of charge. Numerically, e^2 is 10 000 times larger than G_v.

Other, more subtle differences show up in the symmetry properties of weak and electromagnetic interactions. Experiments in the 1950s showed that beta decays are not left–right symmetric; electromagnetic interactions on the other hand preserve left–right symmetry at all times. The lack of left–right symmetry is known as parity violation.

But even in the 1950s there was one apparent similarity between weak and electromagnetic interactions. Both interactions occurred between all sorts of subatomic particle: protons, neutrons, electrons, neutrinos and all the more exotic variants of these that occur only in high-energy cosmic rays from outer space, or in experiments with highly accelerated beams of particles. Furthermore, the weak coupling, G_v, for nuclear decay was measured to be the same as the coupling G_μ for another decay process, that of the muon. The muon is a particle like the electron, only some 210 times heavier. It decays into an electron, a neutrino and an antineutrino (Figure 19.1b). Notice that this decay does not involve particles like protons. So just as electric charge is "universal" in that all particles have charges that are multiples of e, the weak coupling also appears to be universal.

The theoretical explanation of the universality of electric charge is that the electric current of the charged particles involved in an electromagnetic process must be conserved so that charge cannot be created or destroyed. Thus the Soviet physicists S. S. Gershtein and Y. B. Zeldovich in 1955 explained that G_v was equal to G_μ by assuming that there was a similar conserved current between the particles in beta decay and muon decay. This new weak current had also to be electrically charged because in beta decay the neutral neutron turns into a positive proton.

Sheldon Glashow writing his thesis at Harvard University a few years later, and Abdus Salam and John Ward independently in Imperial College, London, took this similarity seriously. They tried to reconcile the different strengths, ranges and parity properties of the weak and electromagnetic interactions, and to convert the two separate theories into a unified theory. They knew that the electromagnetic interaction takes place via the exchange of a photon (Figure 19.2). To make the weak interactions "look" the same, they introduced new particles W^+ and W^-. These two particles would couple to the weak current in the same way that the photon couples to the electromagnetic current (Figure 19.3). Standard rules of quantum theory show then that $G_v = G_\mu = g^2/8M_w^2$, where g is the new weak charge, analogous to electric charge, and M_w is the mass of the W particle.

Glashow, Salam and Ward were thus able to show in 1959 that taking g equal to the electric charge e gave the mass M_w in terms of the two fundamental constants e and G_v, in the same way that Maxwell demonstrated that the velocity of light was given in terms of the two fundamental constants of electricity and magnetism.

Their observation reconciled the differing strengths and ranges of weak and electromagnetic interactions. The coupling constant in both theories was just the electric charge, e; weak interactions were weak and of short range simply because M_w turns out to be so large.

Glashow then showed in 1961 that the parity properties of weak and electromagnetic interactions could also be reconciled

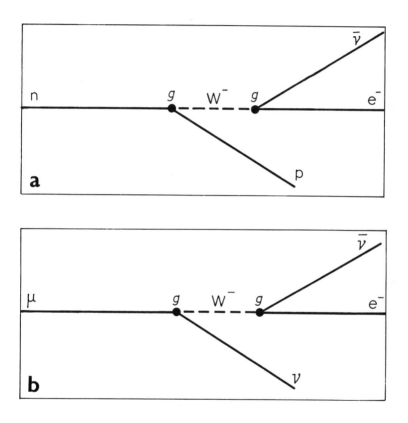

Figure 19.3 *In unified theories, beta decay (a) and muon decay (b) both proceed via the exchange of a W particle, introduced to make these interactions "look" the same as the electromagnetic interaction (Figure 19.2).*

provided that there was a neutral counterpart, W°, to the W^+ and the W^-. He assumed in addition that the W° did not exist in its own right. He argued that "quantum mixing" of the W° and the photon would take place because both neutral particles had identical quantum numbers. A new particle Z° resulted from this mixing mechanism in place of the W°. Glashow was then able to predict that a neutral weak interaction exists due to the exchange of the Z°, for example in electron–proton scattering (Figure 19.4), and that this process would violate parity in a definite way depending on the amount of mixing.

Many experiments since 1973 have confirmed the existence of weak neutral currents (see Chapter 21) and in 1979 Glashow, Salam and Steven Weinberg (now at the University of Texas at Austin) received the Nobel prize for their work on the standard model. At the time even they were surprised at the honour.

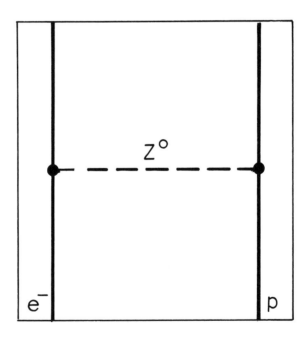

Figure 19.4 *Sheldon Glashow predicted that in addition to W exchange, there should be weak neutral interactions via the exchange of a Z^0 particle.*

Glashow, for example, was reported to say that the Nobel committee took a "bit of a chance" because "nobody has yet built a machine that is capable to check" the new particles predicted (*International Herald Tribune*, 16 October 1979). But that too has changed, as Chapter 22 reveals.

17 February 1983

20

Joining forces in electroweak theory

PAUL DAVIES

A theory that links the weak nuclear force with electromagnetism solves problems that arise in theories of the weak force alone.

Given the bewildering variety of natural phenomena, it is remarkable that, at the subatomic level, all the forces of nature can be explained in terms of just four fundamental interactions. Two of these are familiar in daily life – electromagnetism and gravity. The other two, the so-called weak and strong nuclear forces, are very short-ranged and tend to be restricted to the confines of atomic nuclei. There has long been a hope among physicists that these four forces can be further reduced in number, perhaps to a single, unified force, thus enabling all processes to be described in terms of one truly elementary interaction.

In recent years, physicists have come to realise that the forces that act between material particles at the subatomic level are inextricably mixed up with the structure of those particles themselves. This has led to the use of what is known as the quantum theory of fields as a description of the forces of nature on a microscopic scale.

Until a decade ago, only the electromagnetic force possessed a sensible description in terms of quantum fields, in a theory which explains the electromagnetic action between charged particles as due to the exchange of photons. Thus, for example, an electron will scatter from another electron, or from a proton, as a result of unseen, or *virtual*, photons that are transferred between them.

A central feature of this scheme is that the "messenger" photons exist only fleetingly, by borrowing the energy necessary to their existence for a short duration. When the loan is repaid, the photon disappears. To communicate a force to another particle, a virtual

Plate 20.1 *The initial step in the chain reaction that fuels the Sun is a weak interaction in which a proton turns into a neutron, emitting a positron and a neutrino as it does so. Thus the strength of the weak force controls the rate at which the Sun burns, and ultimately the development of life on Earth. (Credit: RAS)*

photon must live long enough to travel there. But the more energy
the photon borrows, the quicker the debt must be repaid, so only
low-energy, "weak" photons can travel far from the charged
particle. That is why the force between particles diminishes with
distance.

On the other hand, a virtual photon can always be reabsorbed
by the *same* particle. That way it has no distance at all to travel, so
it can be as energetic as it likes. This possibility of "self-action"
with unlimited energy is manifested mathematically in the theory
by the appearance of infinite quantities. At first sight this appears
to reduce the theory to absurdity. However, by careful manipula-
tion, it is possible to absorb the infinities into unobservable
quantities in such a way that the predictions of the theory are
unimpaired; all measurable quantities have sensible, finite values.

This technique of hiding the mathematically undesirable
infinities is called *renormalisation*, and the resulting theory of
quantum electrodynamics (QED), properly renormalised,turns out
to be brilliantly successful (see Chapter 18). The infinities that
arise in QED are symptomatic of nearly all quantum field theories
so that when physicists try to formulate theories of the other three
forces of nature, the same mathematical problems occur. Unfortunately, the renormalisation of QED is by no means automatic, and
only very recently has the technique begun to work for the other
types of interactions.

The strong nuclear force, although unobservable on a macroscopic scale, is the most powerful of the forces of nature. Among
other things it is responsible for binding the atomic nucleus
together against the intensely disrupting electric repulsion of its
constituent charged protons. This was the subject of Part Four,
here we will concentrate on developments concerning the weak
nuclear force.

The weak force operates in a variety of radioactive transmutations, the most familiar of which is the so-called beta decay of free
neutrons, which occurs about a quarter of an hour after their
liberation from nuclei. In this process a neutron suddenly disintegrates, to be replaced by a proton, an electron and an anti-
neutrino – the anti-matter counterpart of the neutrino. The
involvement of neutrinos and their anti-particles is characteristic
of the weak force. These elusive, ephemeral particles interact so
weakly with matter that they can easily pass through the Earth
unimpeded. Nevertheless, in the superdense interiors of some old

tars, a sudden outward eruption of neutrinos from the core can muster enough clout to blast the star to pieces in a titanic cosmic explosion known as a supernova. In the 1950s a number of theorists began to develop a mathematical model of the weak force based on analogy to QED. Once again particles such as electrons, protons and neutrons are envisaged as interacting by exchanging a field particle, this time known as the "intermediate boson" or W particle. An example of this mechanism is depicted in Figure 20.1a, where, in this analogy to QED, a neutron emits a negatively charged W particle, and changes into a proton. The W is absorbed by a nearby neutrino, which transmutes into an electron as a result. (If the incoming neutrino were replaced by an outgoing anti-neutrino, this process would describe the beta decay of the neutron.) Because the participating particles can alter their identity via the weak interaction, the situation is more complicated than in QED.

All experiments prior to 1983 showed only the incoming and outgoing particles; the W was never detected directly. The reason for this is that the W is extremely massive – dozens of times heavier than a proton – so enormous energies are required to liberate a W (see Chapter 22). And the W must pay back the energy loan used to buy its ponderous mass in a time so short that it barely travels a detectable distance before disappearing.

The qualitative value of this picture is obvious, but how successful is it in explaining weak interaction processes quantitatively? Fortunately, the weak force is even weaker than electromagnetism, so it can often be regarded as just a small disturbance, and a series of successively feebler processes treated approximately. Simple processes, such as that shown in Figure 20.1a, are described accurately by the theory (and its recent developments), but more elaborate processes, such as the one depicted in Figure 20.1b, run into trouble.

This latter process appears to the experimenter as the scattering of a neutrino by a neutron, a phenomenon that is exceedingly weak. The theory, however, predicts an infinite effect! Unlike the infinities that arise in QED, those that occur in the W theory cannot be removed by the technique of renormalisation, because the mathematical structure of the theory lacks the vital internal symmetries necessary for the success of this manoeuvre. It wasn't until comparatively recently that a way round this impasse was

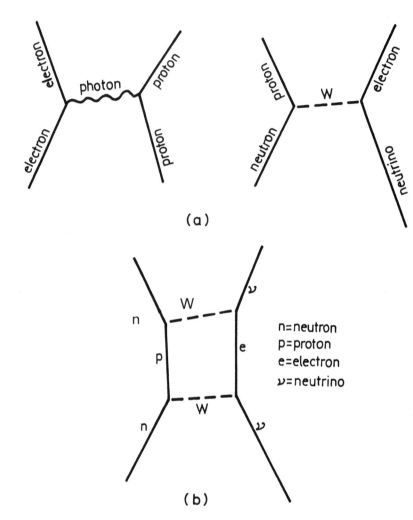

Figure 20.1 *(a) In analogy with QED, where photons are exchanged between charged particles, we can devise a theory for the weak force, in which a heavy, charged particle, W, is exchanged in simple processes such as neutron decay. But more elaborate processes in this kind of theory, such as (b), run into trouble and give infinite answers to calculations.*

discovered. One route to a renormalisable theory is to incorporate what is called *gauge symmetry*. Only mathematics can provide a proper description of this concept, but a simple example conveys a very crude idea.

When an object is lifted from the ground, energy is expended to overcome gravity. This energy can be recovered by releasing the object and letting it fall back to the ground. In the elevated position, the object stores energy internally, as "potential" energy, and its magnitude is given by the simple formula mgh, where m is the mass of the object, g is the acceleration of gravity and h is the vertical height lifted. However, this formula does not actually give the total energy content of the object, only the *difference* between the ground and the elevated location. Because gravity conserves energy, it does not matter by which path the object is transported upwards: only the energy difference between the beginning and end of the path is observable.

The freedom to choose any path, or add any fixed quantity of total energy to the system, without changing the energy difference between the two end-points, is an abstract mathematical symmetry, called a *gauge symmetry* because we can "regauge" our datum of energy without affecting the physical results.

Similar considerations apply to charged particles under the influence of static electric fields. Complications occur if the electric field varies with time or if there are magnetic fields too, but an extended gauge symmetry exists which encompasses all this. Gauge symmetry can be used to deduce all the properties of electromagnetism and is, in particular, deeply associated with the photon's lack of mass. So it seems natural to search for a still more extended gauge symmetry that incorporates both the photon and the W particle.

In such a theory the weak interaction would be treated as simply one part of a more comprehensive electroweak force, the other part being electromagnetism. This step would extend the unification process begun by James Clerk Maxwell who, in the 19th century, showed that electricity and magnetism are really just two interwoven components of a unified electromagnetic force. There is, however, a fundamental problem about extending gauge symmetry to incorporate the weak interaction. Unlike the photon, the W is not massless, and taken at face value, the mass of the W appears to destroy the gauge symmetry and, along with it, the renormalisability of the field theory.

In the 1960s, Steven Weinberg of Harvard University and Abdus Salam of Imperial College, London, discovered a way to circumvent this problem. So long as the mass of the W enters the theory at a fundamental level, renormalisability seems hopeless. On the other hand, if the W could "start out" massless, and acquire its mass as a result of some more superficial mechanism, then the infinities of the theory might become tractable, and renormalisation work once more.

In the Weinberg–Salam theory, the gauge symmetry is not broken in the underlying physical laws that govern the dynamics of the quantum fields. Instead, it is only broken "spontaneously". That is, the W acquires a mass only as a result of the particular (quantum) state it happens to occupy, not as an intrinsic attribute. This subtle distinction can be clarified with the help of an analogy with a more familiar kind of symmetry.

Hold a stick exactly vertically and release it, and it will be in a state of unstable equilibrium – any small disturbance will send the

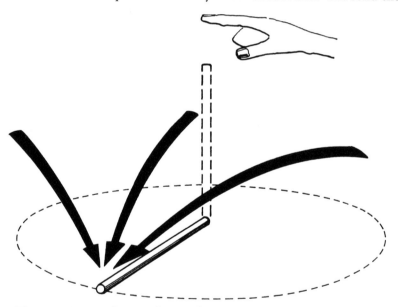

Figure 20.2 *If a stick is allowed to fall from a vertical position, it comes to rest in a particular direction – the symmetry of the original situation has been "spontaneously broken". Symmetry-breaking allows the weak and electromagnetic interactions to be united in the same theory, despite their very different strengths.*

stick toppling to a state of minimum energy, which is a horizontal configuration (Figure 20.2). Because gravity is rotationally symmetric in the horizontal plane, it exerts no bias over the direction in which the stick ends up lying, yet the fact that the stick "selects" one such direction obviously breaks the rotational symmetry spontaneously. The symmetry is still present in the underlying dynamical laws and in the basic structure of the forces, but not in the chosen equilibrium state.

Although these considerations might appear like nit-picking, they turn out to be crucial for renormalisability. In 1971 Gerhart 't Hooft of the University of Utrecht proved that the Weinberg–Salam theory is indeed renormalisable. Since then, several pieces of experimental evidence have supported the Weinberg–Salam model. These include effects caused by the exchange of an electrically neutral Z particle that exists alongside the W in the unified theory, and also a certain lop-sidedness in the electromagnetic scattering of electrons and protons previously restricted to purely weak force processes such as radioactivity.

The success in curing the non-renormalisability of the weak force by unifying it with electromagnetism in an extended gauge theory encourages the search for a still larger gauge theory in which the strong, and perhaps gravitational forces are included.

9 August 1979

21

The discovery of neutral currents

FRED BULLOCK

Tracks in a gargantuan bubble chamber provide evidence for the electroweak theories.

The ultimate aim of research in elementary particle physics must surely be the discovery of some unique principle from which all the laws of nature can eventually be derived. Such a principle has a clear aesthetic appeal in that it gives coherence to all physical behaviour. For the physicist it represents a Nirvana in which all physical processes can be understood and perhaps their effects exactly calculated. This aim is unlikely to be achieved in the near future, if ever. Nevertheless, it seems that progress in the right direction is slowly made. The discovery in 1973 of "neutral currents" at Cern, Europe's international centre for research in particle physics, provide for the first time an indication that some underlying basic principle may indeed exist.

So far as our present knowledge of the Universe goes, it appears that four fundamental types of force (interaction) are required to explain (but not necessarily to quantify) all known physical processes. They are

gravitational	10^{-34}
weak	10^{-10}
electromagnetic	10
strong	10^3

The numbers represent the relative strengths and correspond roughly to the force in Newtons between two protons separated by 10^{-15} metres.

The gravitational and electromagnetic interactions are responsible for familiar macroscopic effects. The former governs the relative movements of massive bodies such as stars and planets, but its influence is negligible in elementary particle physics because the masses of the interacting particles are so minute (for instance, the mass of a proton is 1.7×10^{-27} kg). The second of the two familiar interactions accounts for all phenomena concerned with the propagation of electromagnetic waves, and for the interaction between elementary particles in those cases where the dominating force is that due to the electric charge on the particles. The strong and weak interactions are more obscure in terms of everyday experience but no less important. The strong interaction is responsible for binding together the constituents (neutrons and protons) of the atomic nucleus. It is clearly more effective than the electromagnetic repulsion between the positively charged protons which tends to disrupt the nucleus. The role of the weak inter-action seems to be simply to provide a limiting mechanism on the stability of elementary particles. Occasionally a particle will spontaneously decay into other particles with the release of energy. Such decay processes occur at rates which are very slow by comparison with other phenomena governed by electromagnetic and strong forces. It is rather as though nature has provided, but seldom uses, a mechanism whereby all matter can eventually decay to some stable state.

In view of the immense range (37 orders of magnitude) covered, and the rather clear separation between, the strengths of the four interactions it is perhaps not surprising that attempts to find a fundamental underlying structure have consistently met with failure. The importance of the discovery of neutral currents at CERN was that it provided dramatic confirmation of a unified field theory which describes the weak and electromagnetic inter-actions as different aspects of the same force law. The new theory was, in fact, already some ten years old. Originally discussed by Steven Weinberg of MIT, and by Abdus Salam and John Ward, then at Imperial College, London, it remained in obscurity for some years until Gerhard t'Hooft of Utrecht and Benjamin W. Lee of Stonybrook, New York, showed that it could be "renormal-ised". This is simply the process of proving that a theory gives physically meaningful predictions in all possible applications and never violates basic principles such as, for instance, the conserva-tion of probability. Conventional weak interaction theories have always demanded rather exceptional assumptions to achieve renormalisation. But once a proof of renormalisability is estab-lished it is necessary to take seriously the experimental conse-quences of the theory.

One clear prediction of the unified "electroweak" theory is that the weak interaction should manifest itself not just through so-called "charged current" processes, as had hitherto been exclusively observed, but also through "neutral current" processes. To understand the meaning of "charged" and "neutral current" processes, it is necessary to make a few more remarks about the weak interactions. Radioactive decay is the most basic example of a weak interaction process. Occasionally, one of the protons inside an atomic nucleus spontaneously disintegrates into a neutron, positron (anti-particle of the electron) and a neutrino:

$$p \rightarrow n + e^+ + \nu$$

The apparent non-conservation of energy (the neutron is heavier than the proton) is taken care of by energy drawn from the nuclear field. This reaction contains particles from two different families. On the one hand the neutron and proton belong to a family called *hadrons* (which also includes π-mesons, K-mesons and many others) whose members interact predominantly by means of the strong force. The effects of electromagnetic and weak forces on this family are usually negligible and processes such as those above, occur rather infrequently. On the other hand the electron, neutrino and their anti-particles belong to a family called *leptons* (which also includes the muon) whose members never interact by means of the strong force. An important feature of the lepton family is that its members are always conserved. Thus, if there is one lepton going into an interaction there must also be one, and only one, emerging. In the radioactive decay reaction above, this law is satisfied because no leptons appear on the left-hand side of the equation and on the right-hand side there appears a lepton–anti-lepton pair.

The essential observational difference between leptons and hadrons can be illustrated by considering their behaviour on projection into a thick block of dense material. On entry, both particles would be confronted by a sea of hadrons (the neutrons and protons in the atomic nuclei). The incoming hadron would very soon pass close enough to another hadron to undergo strong interaction and would not then emerge from the other side of the block. The lepton, on the other hand, would have no possibility of interacting strongly with the hadrons and could go a very long distance before the occasional weak interaction would occur. The probability of emergence from the block would be very high. This picture for leptons is not completely correct since, if the lepton is charged, it will undergo frequent electromagnetic interactions with the other charged particles, thereby losing energy and being eventually brought to rest. The relative range of the hadron and charged lepton would be a measure of the relative strengths of the strong and electromagnetic interactions. In practice the large difference in range between leptons and hadrons in matter provides a convenient experimental way of distinguishing unambiguously leptons from hadrons. Consider now the progress of a neutrino in similar circumstances. The neutrino has essentially no mass, it has no charge and does not undergo strong interactions. The only mechanism left to it is therefore a rare interaction through the weak force. A neutrino could traverse the diameter of the Earth 10^{11} times and still have a 50 per cent chance of not interacting! Obviously, if neutrino interactions could be observed, a very powerful means of studying the weak interaction would be available. Modern particle accelerators, such as the proton synchrotron at CERN, can produce bursts of up to 10^{10} high-energy neutrinos or anti-neutrinos every two seconds. With a big enough detector it becomes possible to observe some neutrino interactions and to study their characteristics. Prior to 1973, the neutrino interactions seen were basically of the form

$$\nu + n \rightarrow \mu^- + p$$

although, because of the high energy, the more usual process was

$$\nu + n \rightarrow \mu^- + \text{several hadrons.}$$

The common characteristic of this type of interaction, and of all previously studied examples of the weak interaction involving leptons, is that on emergence from the reaction the ingoing lepton has been transformed to another lepton and in the process the electric charge on the lepton has been changed. A completely different situation exists in the case of the electromagnetic interaction where a positron can interact with a proton

$$e^+ + p \rightarrow e^+ + p$$

and scatter according to the repulsion of the charges but still emerge as a positron. The charge on the lepton has not changed. A physicist's way of looking at such processes is to visualise the leptonic and hadronic components as forming two currents which interact. The leptonic current in the weak interaction is then a charge-changing current ("charged current") while in the electromagnetic it is charge non-changing ("neutral current").

The whole of the weak interaction theory has hitherto assumed that the charged current process is an inherent feature of the weak interaction, and is, moreover, one of the distinguishing features between weak and electromagnetic interactions. Now in the electroweak theory this distinguishing feature disappears and one expects the weak interaction also to proceed via a neutral current process. Thus the theory predicts the existence of processes like

$$\nu + n \rightarrow \nu + \text{several hadrons.}$$

It is precisely this process that was observed in the experiments at CERN in 1973, providing a remarkable confirmation of the theory which is made even stronger by the fact that the process occurs at approximately the predicted rate.

The crucial experiment exploits the observational difference I mentioned above, between lepton and hadrons. An intense beam of neutrinos is produced from the CERN accelerator in the following way. High-energy protons are extracted from the proton synchrotron and allowed to impinge on a heavy metal target. Because the protons interact strongly with the protons and neutrons in the target a large number of hadrons is produced, and these continue along the general direction of the beam. This torrent of particles contains a large flux of π-mesons and K-mesons, many of which decay into muons and neutrinos on the journey through the so-called drift space:

$$\pi^+/K^+ \rightarrow \mu^+ + \nu$$

The drift space is terminated by a monumental wall containing several thousand tonnes of iron and concrete which extracts all the remaining hadrons, through their attenuation by successive strong interactions, and finally all the mesons by electromagnetic energy loss. Emerging from this shielding wall there are then only neutrinos. These traverse the giant bubble chamber Gargamelle, almost five metres long and two metres in diameter, containing about ten tonnes of heavy liquid (freon, CF_3BR). The bubble chamber reveals the passage of electrically charged particles as strings of tiny bubbles in the liquid, the effect being somewhat analogous to the observation of a high-flying aircraft by means of its condensation trail. Although the aircraft itself cannot be directly observed, its passage and exact movements are easily detected because its trail contains all the necessary information. By photographing the strings of bubbles (tracks) one can see exactly how many charged particles have been produced, where they went, and often identify the nature of the particles.

Since the bubble chamber yields a set of stereoscopic photographs every time the accelerator pulses, and almost a million pulses are required to capture a reasonable number of neutrino interactions in the chamber, it is clear that such an experiment in its totality requires the mobilisation of a large effort. The experiment at CERN was carried out by an international collaboration involving physicists from laboratories in Aachen, Brussels, CERN, Ecole Polytechnique (Paris), Orsay (Paris), Milan and University College, London.

In the course of examining the interactions photographed in the bubble chamber it became clear that, while the expected processes giving a charged muon were indeed taking place there was, nevertheless, a significant number of interactions where no charged muon appeared among the interaction products. At a symposium in 1973 in Bonn, West Germany, the Gargamelle collaboration reported 165 muonless events in a data sample containing 576 normal neutrino interactions. Plate 21.2 shows an example of a muonless interaction. It is quite certain that no

Plate 21.1   *A view inside the Gargamelle bubble chamber at CERN. (Credit: CERN)*

charged lepton is produced since all the particles interact within a relatively short distance from production whereas a muon would have left the chamber without interacting. The many examples of normal charged current processes observed are all clearly distinguished by the presence of a particle which travels a large distance in the liquid before eventually leaving the chamber. The inference is that the muonless interactions are manifestations of the neutral current process where the lepton coming out of the interaction is a neutrino which cannot, of course, be detected.

Before claiming these interactions as examples of neutral current processes it was necessary to be sure that no other known process could produce interactions with the observed characteristics. In this particular case, a potentially embarrassing background exists. Neutrinos interact in the shielding wall as well as in the bubble chamber. Among the interaction products are neutrons which can emerge from the shielding if the parent neutrino interaction is close enough to the end of the wall. Neutrons are uncharged and leave no tracks in the bubble chamber but soon undergo strong interaction to produce charged hadrons giving events in the chamber which can look topologically identical to neutral current interactions.

Fortunately, Gargamelle is so big that neutrino interactions occurring in the chamber frequently give a neutron which then reveals its presence by interaction also in the chamber. By studying the characteristics and frequency of neutron production from neutrino interactions it was possible to show that background from neutrons could occur at a rate no greater than 4 per cent in neutrino events, and 8 per cent in anti-neutrino events, of the normal charged-current interaction rate. The muonless events occurred at a rate of 23 per cent and 46 per cent of that for the charged current reactions of neutrino and anti-neutrino respectively. The signal to background ratio was therefore a comfortable 6 to 1. Moreover, the distribution of muonless events along the chamber was observed to be more or less uniform. If the events were due to neutrons they should have occurred predominantly at the front end of the chamber due to the strongly interacting characteristics of the neutron.

The team made many other checks on the events, all of which showed that the characteristics of the hadrons produced are in every way identical to those of the hadrons produced in normal charged-current interactions. The evidence was therefore overwhelming that some hitherto unobserved process was occurring

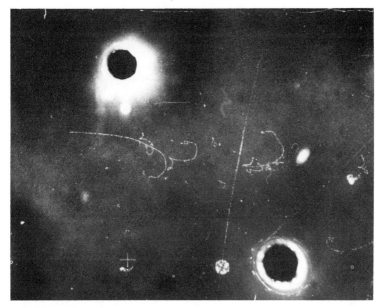

Plate 21.2   *An (invisible) neutrino enters the Gargamelle bubble chamber from the left and scatters from an electron in the liquid which then moves forward. This is one of the first observations of an interaction occurring through the agency of a weak neutral current, demanded by theories that unite weak and electromagnetic interactions. (Credit: CERN)*

which in all respects showed a neutrino-like origin. It was gratifying, even surprising at an early stage, that the observed events occurred at much the rate required by the electroweak theory. It was also gratifying that an experiment of similar type (but at higher neutrino energy and using different detection methods) mounted at the Fermi National Accelerator Laboratory in the United States, soon began to produce data which confirmed beautifully the European results.

Some 18 months before the discovery of neutral currents Weinberg said of his new theory – "Right now there's not a grain of experimental evidence that this general idea is right. But it solves so many theoretical problems all at once that it smells right." It appears that Weinberg's olfactory sense was functioning with exceptional efficiency.

4 October 1973

22

The quest for the W and Z particles

CHRISTINE SUTTON

High-energy collisions between protons and anti-protons reveal the long-sought W and Z particles, the carriers of the weak force that are demanded by electroweak theories.

An atmosphere of excitement mingled with a fair amount of caution and some controversy took over CERN, Europe's centre near Geneva for research into subatomic particles, in the first few weeks of 1983. In seminars held there towards the end of January two groups of scientists announced the first evidence of CERN's greatest discovery yet, a particle known simply by the letter W. The discovery vindicated not only all those who pushed for the means to make it possible. It also indicated that theorists are following the right path in their attempts to unite the disparate forces of nature within one mathematical framework.

Why should "yet another" subatomic particle cause so much excitement? The story really begins in 1935 when Hideki Yukawa, a Japanese theorist, put forward the idea that the forces operating in the atomic nucleus should have associated with them specific "exchange" particles. These would flit between the protons and neutrons in the nucleus "carrying" the nuclear forces. There was already a strong conviction that the electromagnetic force acted in this way, carried by photons ("packets of light") passing between electrically-charged particles. Thus Yukawa proposed that the strong nuclear force, which holds protons and neutrons together, was carried by particles later named "mesons". These mesons were eventually discovered in experiments with cosmic rays – energetic particles from outer space – in 1947. Yukawa also suggested that a similar particle must mediate the weak nuclear force, which underlies many forms of radioactive decay.

The weak force is some thousand times weaker than the electromagnetic force, and 100 000 times more feeble than the strong nuclear force. But like the strong force, and unlike the electromagnetic force, the weak force has a finite range: it acts only over distances of the size of a nucleus, about 10^{-13} cm. The feeble nature of the weak force and the short interaction distance together imply something about the nature of the particle exchanged: it must be relatively heavy. The particle being exchanged in a sense "borrows" its mass from the force field; the heavier the mass, the shorter the time it can be borrowed, and, therefore, the shorter the range of the force. (The photons that carry the electromagnetic force have no mass; this corresponds to the force having infinite range, as experiments show to be the case.) Yukawa's meson, the transmitter of the strong force, turned out to have a mass roughly 15 per cent that of the proton and some 270 times that of the electron. However, even simple estimates for the W particle suggest that it should weigh in at something like 30 times the proton's mass.

In the late 1960s theorists began to acquire a more precise concept of the role of the W particle in weak interactions, and could make better predictions about its mass. In particular they made great progress in their attempts to put electromagnetism and the weak nuclear force together within the same theoretical framework. It turned out that a mathematically respectable "electroweak" theory required three exchange particles for the weak force: two electrically-charged particles, W^+ and W^-, and a third neutral partner which has become known as the Z^0. An important boost for this theory came in 1973 when experiments at CERN discovered the so-called neutral currents: weak interactions between particles that occur via the exchange of something neutral, in other words the Z^0. Over the years more and more evidence consistent with the electroweak theory has come from numerous experiments not only at CERN but at particle accelerators throughout the world, and in 1979, Sheldon Glashow, Abdus Salam and Steven Weinberg received the Nobel prize in recognition of their theoretical endeavours to unite the weak and electromagnetic forces. But still there was no explicit experimental evidence that the W and Z particles, key figures in the theory, did indeed exist.

According to electroweak theory, and using results from these other experiments, the W particle should have a mass of 80 gigaelectronvolts (GeV), or 80 times the mass of the proton. This in itself is sufficient to explain why the particles have not been observed previously. To create this mass, rather than "borrow" it as in the exchange, requires an equivalent amount of energy. In the mid-1970s there already existed accelerators, both at CERN and at Fermilab in the US, that could give protons with energies up to 450 GeV. At first sight this energy seems sufficient to produce "free" W particles copiously. But the high-energy protons from these machines are directed at targets containing stationary protons and much of the energy of the bombarding particles is used up in giving kinetic energy to particles of relatively low mass – protons, mesons and so on – created in the collision. The energy left over, about 30 GeV, is not enough to create W particles, if electroweak theory is correct.

One way round this apparent impasse is to collide together particles with equal amounts of energy travelling in opposite directions; then all the energy is available to create new particles. And Carlo Rubbia, from CERN, along with David Cline, from the University of Wisconsin, and Peter McIntyre proposed an ingenious way to convert existing accelerators such as the 450 GeV Super Proton Synchroton (SPS) at CERN, to operate as particle colliders. Their idea was to feed anti-protons, like protons but oppositely charged, into the main magnet ring of the proton accelerator. The anti-protons would follow the same path through the magnets – but in the opposite direction to the protons – and would be accelerated to the same energies. Then at the right moment the two beams of particles could be brought together to collide.

Convinced by Rubbia's dynamism and unique art of persuasion, CERN agreed in 1978 to convert the SPS to a proton–anti-proton collider (see next article). The plan was to install experiments at two of the points at which protons and anti-protons would collide on their journey round the accelerator's ring of magnets. The main experiment at one point, code-named UA1, has been designed under Rubbia's direction by over 120 scientists from Austria, Britain, France, Germany, Italy, the US and CERN. At the other point the primary experiment UA2 has been put together by 50 or so researchers from France, Denmark, Italy, Switzerland and CERN.

Both UA1 and UA2 were designed to be capable of detecting W and Z particles. The two sets of apparatus complement each other to a certain extent and also provide a useful means of cross-checking results, as well as friendly competition.

By July 1981, physicists and engineers at CERN had succeeded

Plate 22.1 *The UA1 experiment at CERN – more than 1000 tonnes of apparatus to search for the W particle in proton–anti-proton collisions. (Credit: C. Sutton)*

in building a suitable source of anti-protons that could feed the SPS. It was then that the first proton–anti-proton collisions at 270 GeV per beam took place. The energy of these head-on collisions is equivalent to that of a beam of 155 000-GeV protons striking a stationary target: surely enough to reveal the W and Z particles if they exist.

From August 1981 onwards scientists on UA1 and UA2 worked at putting the finishing touches to their apparatus, and at the same time machine physicists at CERN tried hard to get a large enough number of collisions in the SPS. Only a few of the particles fed into the SPS actually collide; the remainder just fly past each other. And only in a minute fraction of the actual collisions does theory suggest that a W particle will be produced; many different processes compete to take place each time a proton and anti-proton meet. Coupled with problems that arose with the apparatus – UA1 for example was beset by difficulties ranging from dirty air-lines to burst pipes – it was not until November and December 1982 that the experiments were able to observe a large sample of proton–anti-proton collisions.

At last it seems that the waiting may have borne fruit. In a seminar at CERN on 20 January 1983, Rubbia presented five events, or collisions, in which the particles that emerge are consistent with what is expected for a particular mode of decay of the particle. The following morning, a member of the UA2 team presented four events detected in that apparatus.

How did the experimenters know what to look for? The situation is similar to that for neutron decay, the reaction for which theories of the weak force were first developed. A neutron decays into a proton, electron and neutrino. In this process the neutron transmutes to a proton by emitting a W particle. The W then decays to an electron and a neutrino. What the experimenters at CERN looked for is a proton meeting an anti-proton to produce a W particle which then decays to an electron and a neutrino.

At a more fundamental level theorists describe the creation of the W particle in terms of quarks, the building blocks from which the protons, neutrons and their anti-particles (and many other particles) are made. The proton comprises three quarks; the anti-proton is made of three anti-quarks. One of the quarks and one of the anti-quarks meet and, in a process called annihilation, produce a W particle. This then decays to an electron and a neutrino (Figure 22.1). By generating mesons, the carriers of the strong nuclear force, the remaining "spectator" quarks materialise as swarms of low-energy particles. But the electron carries a high

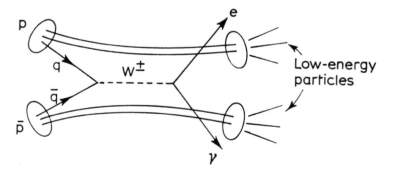

Figure 22.1   *The formation and decay of a W particle in a proton–anti-proton collision. The W is formed in the annihilation of a quark and an anti-quark from the initial particles; it decays into an electron (or a positron) and an undetectable anti-neutrino (or a neutrino). Many other particles emerge at the same time, produced by the interactions of the other quarks and anti-quarks.*

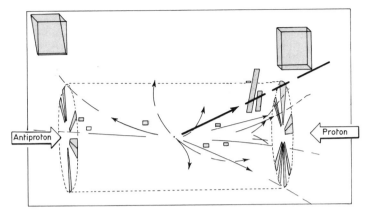

Figure 22.2 *A "W event" in the UA1 apparatus. The heavy line shows the track of the electron in the central cylindrical detector. It deposits lots of energy in the electron detector (shown by the long lozenge shape). Other tracks are from low-energy particles produced by the "spectator" quarks as in Figure 22.1.*

energy transverse to the direction of the collision, and thus bears the signature of the W particle that gave it birth. The neutrino also carries with it a large transverse energy but the particle's properties are such that the experiments at CERN cannot detect it and it escapes the apparatus unseen. The "signal" to look for, then, to observe a W, is a single electron carrying a great deal of energy.

In its preliminary analysis, the UA1 team found five occasions on which a single electron with all the right characteristics was produced. This was out of the 850 000 or so collisions the UA1 apparatus recorded towards the end of 1982 from a total of about one thousand million actual collisions. The UA2 team found four similar "events" out of roughly the same total number of collisions. Are these events, like the one shown in Figure 22.2, really the footprint of the long-awaited W particle?

The events have everything going for them. The apparatus in both UA1 and UA2 cannot specifically identify electrons, but it can eliminate other possibilities. In the case of UA1, for example, a gas-filled cylindrical detector surrounding the collision point senses the tracks of electrically-charged particles as they move through a magnetic field. Wrapped around this is a detector (a calorimeter) designed to collect all the electromagnetic energy carried by electrons and photons. Beyond this a third detector picks up all the energy of hadrons – particles like protons and mesons that are built from quarks.

In the five interesting events the detectors of UA1 show a track in the central detector leading to a single cell in the electromagnetic calorimeter, which picks up a lot of energy. But the track leads to no signal at all in the hadron calorimeter (Figure 22.2). This track, it seems, *must* belong to an energetic electron. Furthermore, calculations on the balance of energy in each event are consistent with there being a missing energetic neutrino. Together the energies of the electron and neutrino give a lower limit of 73 GeV for the mass-energy of the intermediate state that created them, remarkably close to the value that theory predicts for the W particles. And, to make it all seem almost too good to be true, the total number of possible W particles seen by both UA1 and UA2 is just what one would expect from the total number of proton–anti-proton collisions that the two experiments have observed.

27 January 1983

When matter meets anti-matter

The concept of putting anti-protons into a proton accelerator is easy to state; the practicalities are another matter. One of the main difficulties lies with ensuring that there will be sufficient collisions between protons and anti-protons to make the endeavour worthwhile. It's a little like firing two shot-guns at each other: most of the particles will go straight past each other without colliding. Moreover, while protons are relatively easy to come by – strip a hydrogen atom of its single electron and you are left with a proton – anti-protons can be made only in high-energy collisions of other particles.

This process is, however, like panning for gold. A beam of 10^{13} protons with 26 GeV of energy creates only 2.5×10^{7} anti-protons with an average energy of 3.5 GeV when it strikes a target of tungsten. But calculations show that to produce a useful number of interesting interactions in proton–anti-proton collisions at 270 GeV – the energy used in the Super Proton Synchrotron at CERN – you need to collide bunches that each contain at least 10^{11} particles. A further problem arises because the anti-protons selected from the particles emerging from the target have a range of velocities and directions. Once within the confines of a ring of accelerator magnets, many of these anti-protons would stray from the ideal path, which is suited to one velocity, and became lost, thus reducing the total number even further. Somehow the beam of anti-protons must be both intensified and concentrated in some sense, before it enters the main accelerator ring and to this end CERN built a new machine called the Anti-proton Accumulator.

The means for concentrating anti-protons that CERN has chosen is a technique called stochastic cooling, which was first proposed by Simon van der Meer, a physicist at CERN, in 1972.

Plate 22.2 *The ring of the Anti-proton Accumulator at CERN, where anti-protons are stored before transfer to the main ring of the Super Proton Synchrotron. There they are accelerated to 270 GeV and collided with a counter-rotating beam of protons. (Credit: CERN)*

The basic principle of the technique is to sense the average position of a bunch of particles at some position in a ring of magnets, to work out the correction that needs to be made to this value, and to apply the correction at some point further round the ring. This procedure involves sending a signal across the ring before the particles get there – no mean feat when the particles are travelling close to the speed of light. Developing the technique thus relied on designing electronic circuits that respond fast enough to send a signal along a cable across the arc of the ring, so as to beat the particles to the correction point. There the signal stimulates "kicker" magnets (electromagnets, like all those at accelerators) to provide the correct magnetic field to adjust the path of the particles. Successive adjustments concentrate the spread of the particles' velocities.

But "cooling" like this only concentrates the beam. To produce bunches that contain more particles, successive bunches must be stored together. The Anti-proton Accumulator must do this task in addition to cooling the beam. Each bunch of 10^7 anti-protons that enters the accumulator is allowed a few seconds of rapid "precooling" in which its momentum spread is reduced from $\pm$ 7.5 to $\pm$ 1 parts per thousand. The bunch is then shifted to one side in the vacuum tube where it joins the "stack", which contains all previous bunches, and which is usually shielded from the section of the ring where precooling takes place. While circulating in the stack the anti-protons are further cooled, and more bunches are added every 2.4 seconds, until after 24 hours the stack contain some 10^{11} anti-protons – the magic number for useful physics in proton–anti-proton collisions.

20 August 1981

"This week"

CERN, Europe's centre for research into the fundamental nature of matter, may have pulled off a second important discovery in 1983. In May one group of researchers announced that it had found signs of the elementary particle known as the Z^0 (pronounced Z-nought). This news came hot on the heels of the discovery earlier in the year concerning W particles, the electrically charged partners of the neutral Z.

The W and Z particles figure prominently in attempts to develop a single unified theory of the particles that form the matter of the Universe and the forces that mould them together. Generally, different theories are used to explain widely different phenomena, from the forces in the atom to gravity. But unlike the W particles, which can be predicted by other "non-unified" theories, the Z^0 is a feature only of a theory that unifies forces in a particular way. As such the Z^0, more than the W particle, is an indicator that the theorists are on the right track.

The theory that predicts the existence of the W and Z particles evolved largely through efforts to understand the weak nuclear force. This force is effective at distances of one hundredth the diameter of the proton, or 10^{-15} cm. Without this weak force the Universe would be very different, for the force mediates the radioactive decays of many subatomic particles and atomic nuclei. Such weak interactions underlie the processes that fuel the Sun and other stars.

The idea was to develop a theory in which the weak force is carried by a type of particle. The precedent for this approach lies in the successful theory of quantum electrodynamics (QED), which describes the electromagnetic force in terms of the exchange of photons. The photons are the massless packets – "particles" – of energy that constitute light and all other forms of electro-magnetic radiation. QED pictures electrically-charged particles, such as electrons, interacting by exchanging photons, much as rugby players interact by passing a ball between each other.

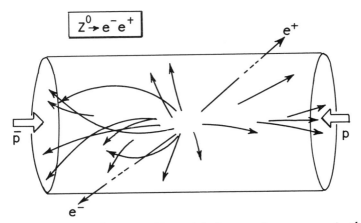

Figure 22.3   *A decaying Z particle betrays its presence in the straight back-to-back tracks due to an energetic electron–positron pair (e^+e^-), in the UA1 apparatus. The tracks are picked up in the gas-filled central detector, which observes the collisions of high-energy protons (p) and anti-protons ($\bar{p}$).*

In the case of the weak interaction the "balls" must be very heavy, because the force is so feeble and has such a short range: it would be difficult to throw a rugby ball made of lead very far, very often. The carrier of the weak force would also have to be electrically charged, for a neutron to decay into a proton, for example. Thus was born the idea of the W particle, which, according to the latest versions of the theory, should weigh in at some 85 times the proton's mass, just as two teams at CERN found earlier in 1983.

A theory with only charged W particles turns out to have infinite quantities that arise in calculating certain effects. But these embarrassing infinities cancel out if the theory includes not only the weak force, but also the electromagnetic force, and two neutral carriers as well as the charged W particle. One of these neutral particles is simply the photon, the carrier of the electromagnetic force. The other is the Z^0, which is predicted to have a mass in the region of 95 times that of the proton.

The group of researchers that reported the first indications of a Z particle is code named UA1, and is one of the two teams that announced evidence for the W particle. UA1 is a collaboration of 100 or so physicists from Europe and the US, led by Alan Astbury from the UK's Rutherford-Appleton Laboratory and Carlo Rubbia from CERN.

The team was scrutinising new data collected during April 1983, when early in May they found one proton–anti-proton collision recorded in their apparatus which bears the hallmarks of a Z^0. What the apparatus "sees" is the decay of the Z^0 into an electron and a positron (an anti-electron) flying off back to back from the point of the collision. If they do really emanate from a Z^0 the electron and positron should together carry energy equal to the particle's mass, and this seems to be the case.

12 May 1983

23

Hidden symmetry and the Higgs particle

ANDREW WATSON

Theories that link the electromagnetic and weak forces are not quite home and dry. There remains the mysterious "Higgs particle" still to be discovered.

Twice in 1983, scientists at CERN, Europe's Geneva-based centre for research in sub-nuclear physics, tentatively announced the discovery of a new particle. The physicists hoped that they had at last observed the W and Z bosons, which are of special significance for they play key roles in our understanding of nature's fundamental forces, or interactions. If these particles did not exist then the theorists of particle physics would be in grave trouble; but even if, as seems likely, the W and Z particles do exist, the theorists are not home and dry. The underlying theory that predicts the existence of these particles must embody certain fundamental symmetries of nature, and in so doing has led to concepts completely new to particle physics, borrowed, as it were, from solid-state physics. To prove conclusively the validity of these concepts in particle physics, more experimental evidence is needed: there are other particles that follow in the steps of the W and Z particles.

The W and Z particles belong to the realm of the weak interaction, one of four fundamental interactions governing the Universe. The weak force is responsible for certain radioactive decays, as well as the nuclear burning process that produces heat from the Sun. In the 1930s Enrico Fermi, then at the University of Rome, developed a theory of the weak force, but this was later shown to violate a well-established physical principle and was for this reason ultimately unsatisfactory. Subsequent attempts to improve on Fermi's theory centred on the introduction of a new particle to act as intermediary in the weak interaction. This new particle we now call the W boson.

Such an adventurous step was not taken in total darkness. Proponents of the early versions of these theories incorporating an intermediary particle, for example Julian Schwinger at Harvard University, based their attempts on a theory that already existed, known as quantum electrodynamics, or QED. This is a description of the electromagnetic interactions of sub-atomic electrically charged particles. These particles, for instance electrons and protons, are understood to interact by exchanging a massless intermediary, the photon, which can loosely be described as a particle of light.

Any theory of physical interactions must respect the conservation laws of those interactions. In the case of electromagnetism, for example, electric charge is conserved. In other words, the total charge always remains the same; if a positive charge is created, then so must be a negative charge, to give a net increase of zero. Moreover, the existence of a conservation law implies a symmetry in our mathematical description of the physical world. As an everyday example of a symmetry, imagine a dinner plate rotated about its centre through any number of degrees. The initial and final positions are indistinguishable: the plate has rotational symmetry. A mathematical symmetry means that we can change our equations in such a way that the initial and final forms describe the same physics.

What kind of symmetry must a theory of physical forces possess? Consider first QED, and the symmetry associated with charge conservation. Charge is always conserved in the immediate vicinity: it is a *local* phenomenon, a principle embodied in the theory of electromagnetism that James Clerk Maxwell developed in the early 1860s. But the mathematical symmetry corresponding to charge conservation is *global*, where global means "happening everywhere at once" (Figure 23.1). Prompted by the observed local nature of charge conservation, we may enforce a similar local principle on our otherwise global mathematical symmetry. This step has profound consequences; it demonstrates exactly how the electromagnetic force between charged particles is to be understood in terms of the exchange of photons.

A theory of this type, where a symmetry corresponding to a conservation law is forced to become local, is called a "gauge theory". The new particles introduced to transmit the force are called "gauge bosons". Thus the photon is the gauge boson of

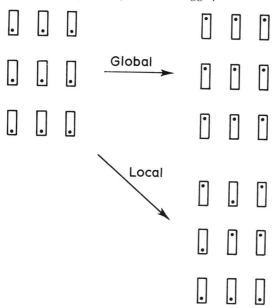

Figure 23.1 *These oblongs are symmetric to rotations of 180°. In a global transformation all the shapes turn simultaneously, as the dots indicate. Under a local transformation, however, each shape exhibits its symmetry independently.*

QED. One aim of theorists since the successful development of QED has been to construct a gauge theory of the weak interactions, where the corresponding gauge boson is the W particle.

Now, unlike the photon, the particle responsible for transmitting the weak force can have positive or negative charge, for it can turn a neutral neutron into a positive proton, as in the process of nuclear beta decay. It must also be massive, to account for the short range of the weak force. An intermediate particle borrows energy from the surrounding particles by an amount proportional to its mass. Borrowed energy must be returned in a short time, according to the rules of quantum mechanics, and the more energy that is borrowed, the sooner it must be repaid. Therefore a heavy particle may borrow its energy for a short time only.

To follow the gauge theory prescription as used in QED, it is necessary to identify conserved quantities of the weak force and hence the underlying symmetry. This turns out to be much more complicated than simple charge conservation and its associated symmetry in QED, as there are several peculiar features of weak interactions that the theory has to accommodate. But in 1961, Sheldon Glashow, at Harvard University, finally devised the correct mathematical symmetry relation for the weak force. Glashow's symmetry grouping could accommodate both charged W bosons, represented by W^+ and W^-, as well as two uncharged particles. One of these was to be identified with the photon, and was the first hint of a mathematical intertwining of weak and electromagnetic theory. There was still a problem, however. Glashow, like Schwinger and others, was unable to include the requisite mass of the W boson in a consistent way.

In fact, gauge theories suffer from a difficulty that I have so far conveniently ignored. The special local symmetries of gauge theories hold good only when the intermediary particle, the gauge boson, is massless. For QED there is no problem: the gauge boson of QED is the massless photon. But the weak force requires a massive gauge boson – the W particle. Here the symmetry would appear to be destroyed, and the gauge theory prescription seems to fail.

In looking for ways to resolve the problem concerning the W's weightiness, physicists investigated other physical systems that could perhaps provide a vital clue. In 1963, P. W. Anderson of the Bell Telephone Laboratories demonstrated that a superconductor in a magnetic field was just such a system.

Applying a magnetic field to a collection of mobile charged particles induces electric currents. These new currents in turn produce their own magnetic field, which serves to oppose the applied field. This is the basic mechanism of diamagnetism in ordinary "non-magnetic" materials. In superconductors, which are perfect diamagnets, the effect becomes quite dramatic.

Certain metals become superconducting when they are cooled below a specific "transition" temperature. Below this temperature, any current initiated in the material continues unabated; the electrical resistance is zero. In the superconducting state, the metal's electrons, responsible for normal conduction at room temperature, become loosely bound together into electron pairs, despite the usual tendency of electrons to repel one another. These so-called "Cooper pairs" behave very differently from normal conduction electrons, and give rise to the superconductor's zero resistance. In 1931, W. Meissner and R. Ochsenfeld working in

Berlin observed a related effect. When they placed a potential superconductor in a magnetic field and lowered the temperature to below the transition value, all trace of magnetic field vanished from the interior of the material. This is because currents induced in the surface of a superconductor are sufficient to screen out the applied field completely.

We could think of the induced field due to the screening as exerting a drag on the applied field, by an amount proportional to the strength of the applied field. As the applied field attempts to penetrate the superconductor, the drag effect increases; the field would appear to be gaining mass as it interacts with the superconductor. In fact, the mathematical description of the way a magnetic field behaves in the screened region is identical to quantum theory's description of a heavy particle as a wave. The field behaves as though it were acquiring mass via the screening effect of the induced currents.

This, then, is the clue to the problem of the W's mass that we were looking for. The concept of a magnetic field acquiring mass may be generalised into the language of particle physics, where it becomes the idea that a gauge boson acquires mass. This sounds hopeful, because the W is a gauge boson, and we want it to be massive.

There exists one important difference between the example of the superconductor and particle physics, however. In the former, screening of the external field is provided by the interaction of that field with the material of the superconductor itself. In particle physics we think of particles moving through empty space, the vacuum. If the example of superconductivity is going to be useful, we need to study the way in which screening currents result in massive gauge bosons. What is to cause the screening current in free space?

The answer lies in reviewing what we mean by free space, or vacuum. Instead of thinking of the vacuum as nothing, that is completely empty, we reinterpret it as a state of minimum energy. Such a state need not be utterly empty, but can supply some sort of background, provided that background minimises energy. This background presence in the particle physics vacuum is provided by a new field called the Higgs field, after its inventor Peter Higgs of the University of Edinburgh. For most fields, energy is a minimum when the field is zero everywhere, clearly a highly symmetric situation. For the Higgs field, however, energy is a minimum when the field has some finite uniform value. The Higgs field will be the source of screening currents in the "vacuum", providing means of bestowing mass on the W boson of the gauge theory of weak interactions. At first sight, this prescription seems totally arbitrary and without foundation. The Higgs field is indeed an invention, and its possible physical origin is subject to debate. But we can turn for justification and guidance to solid-state physics once again, this time to ferromagnetism.

A ferromagnetic material such as iron consists of millions upon millions of spinning atoms, each of which behaves like a microscopic magnet. Above a certain transition temperature, the Curie temperature, these multitudinous atomic magnets are totally without order. However, below the Curie temperature, the atomic magnets are ordered, in other words the magnetic direction of each atom is aligned with those of its neighbours. Thus as the temperature of a ferromagnetic material falls below the Curie temperature, the material becomes spontaneously magnetised. Above the Curie temperature the lowest energy state corresponds to zero magnetisation; below it, minimum energy corresponds to a finite magnetisation.

The way in which the atoms are aligned is of particular interest. All that is required to give a finite magnetisation is that the atoms are aligned; any direction is a potential alignment direction, and all directions are equally good. Both the underlying theory and the possible magnetisation directions have rotational symmetry, like the dinner plate; the aligned system clearly does not. In the process of alignment the atoms have picked out a direction at random, thereby spontaneously destroying, or "breaking", the symmetry. In fact the symmetry is best described as "hidden", as the underlying theory preserves its rotational symmetry throughout, although the physical system loses this property below the Curie temperature. Such symmetry-breaking (or hiding) is a general feature of systems where the lowest energy state – the ground state – is not empty.

How does ferromagnetism help us understand the mechanism of the Higgs field? It provides an example of a physical system with a ground state that is not empty. The magnetisation (below the Curie temperature) is the magnetic model of the Higgs field. The correspondence in ferromagnetism between lowest energy and finite magnetisation becomes a correspondence between lowest energy and a finite Higgs field. The direction of magnetisation, the choice of which was arbitrary and corresponded to breaking symmetry, has an analogous spontaneous symmetry-breaking in the Higgs field. Ferromagnetism illustrates the idea of introducing the Higgs field into the vacuum. And the Higgs field, behaving like the material in a superconductor, provides screening currents in the vacuum, which in turn give mass to the W boson.

In the late 1960s, Abdus Salam of Imperial College, London, and Steven Weinberg at Harvard University developed independently a model of the weak interaction within Glashow's symmetry scheme, but which explicitly included the Higgs mechanism to account for the appearance of massive gauge bosons. The specific proposal of Weinberg and Salam was the introduction of four Higgs fields via a spontaneous symmetry-breaking of the vacuum. The gauge bosons of the theory acquire mass in propagating through this new, asymmetric vacuum. Three of the Higgs fields lend mass to three gauge bosons. Two of these are the W^+ and W^-; the third is another heavy particle denoted Z^0, which mediates a whole new family of weak processes, subsequently discovered at CERN in 1973. The fourth gauge boson is the massless photon, and the fourth Higgs field should give rise to an observable massive "scalar" Higgs particle.

For their work on the Glashow–Salam–Weinberg (GSW) theory, all three were jointly awarded the 1979 Nobel prize in physics. In one go the theory not only provides a theory of the weak force, but mathematically links it with the electromagnetic force. It also makes some very rigorous predictions.

There are few unknown quantities in the theory, as a direct consequence of the Higgs mechanism. So, measurements of radioactive decays allow predictions of the W^+, W^- and Z^0 masses, and their interaction strengths, to within very stringent limits. This is why the observation of the W particle is so important; not only must it be there, but it must have exactly the predicted mass, just as the researchers at CERN have found.

Experimental physicists always hoped to see the W and Z^0 particles first. These are perhaps the most important tests of the GSW theory. But what of the Higgs particle? The Higgs mechanism, which made all the above predictions possible, also predicts the existence of a new massive particle.

Unfortunately we do not know all we would like to about this particle. We know it is massive, has zero charge and no intrinsic spin. A particular problem concerns its mass. Without a prediction of the mass, the experimenters do not know where to look.

Plate 23.1 *In 1979, Sheldon Glashow (left), Abdus Salam (centre) and Steven Weinberg (right) shared the Nobel prize for physics for their work on a unified theory of weak and electromagnetic interactions. Their theory requires the existence of the W and Z particles, and a particle called the Higgs, in order to be finite and consistent. The W and Z have been found, but where is the Higgs? (Credit: Reportagebild/Tobbe Gustavsson)*

A number of theorists have invoked an assortment of physical principles in attempts to estimate limits on the mass of the Higgs particle. But the results of these calculations demonstrate such disparity that the problem of the Higgs mass remains a very open question. A more complete description of the Higgs particle may only become possible with some extra theoretical ideas. Currently there is some hope that the so-called supersymmetric theories (see Chapter 26), which "unify" matter particles with gauge bosons, may provide a more natural explanation of the Higgs, as a relative of some already known particle if the Higgs particle really does exist and has quite a low mass, it could be produced by decaying W or Z particles. In this case it may materialise at CERN, in the same experiments that appear to have successfully created the W and Z.

Despite minimal experimental evidence, the Higgs mechanism has become widely applied within particle physics. It is used in systems where a physical manifestation has less symmetry than the underlying model, or where mass must be bestowed on otherwise massless particles. For example, recognising that the GSW theory is by no means the end of the unification story, many physicists are currently engaged in the construction of so-called grand unified theories (see Chapter 24). The hope is to find a master symmetry that somehow embodies all the forces of nature. This symmetry may have been exact at some early stage in the growth of the Universe, or perhaps at a huge energy. For everyday energies, however, the symmetry is broken. In these theories the Higgs mechanism is used successively to break down the master symmetry into distinct symmetries or distinct forces as we observe them in the laboratory. Needless to say, certain particles acquire mass in the process.

There are many competing grand unified theories, and other ideas which all link particle physics firmly with cosmology. Observation of the W and Z tests the unification of weak and electromagnetic forces. Observation of the Higgs particle may hint at the next link in the chain, and put the interweaving of the nature and formation of matter on a firmer footing.

26 May 1983

PART SIX
A window on creation

For many years the holy grail of physics has been the single, all-embracing theory that unifies nature's rich variety. A variation on this theme was manifest in the 19th century, for example, with the feeling that physicists were close to having all the answers. Then came the discovery of radioactivity and the opening up of the mysterious quantum world inside the atom. Now theorists have come to terms with quantum phenomena, and are attempting to develop theories that can encompass all the workings of the physical Universe, and thus, at least in principle, provide all the answers. Strange as it may seem, the new worlds opened up by the discovery of radioactivity, have in fact brought physicists closer to their goal of total unification. It now seems that the more we learn about the behaviour of sub-atomic particles and the forces that control them, the more we learn about the Universe as a whole, and in particular how it may have been much earlier in its history.

Part Five has shown the success physicists have had in uniting the weak and electromagnetic forces in a single theory. In Chapter 24 Sheldon Glashow (who shared the Nobel prize for his part in developing the electroweak theory) discusses how these ideas can be extended to incorporate the strong force in the so-called "grand unified theories". He also allows us a glimpse of how particle physics can provide a window on creation. The hot big bang with which the Universe is commonly believed to have begun, provides a "laboratory" with far higher energies than physicists can ever hope to achieve with accelerators. In these first instants, the strong, weak and electromagnetic forces would play equal roles. It is only as the Universe has cooled, that each force has one by one "crystallised out" to give the very different interactions we observe today.

Fortunately the grand unified theories do make predictions that we can test in our present Universe, and so provide us with a "telescope" with which we can in a sense view the early Universe. For example, Glashow describes how massive monopoles – particles with single magnetic poles – should have formed copiously in the early Universe. In the two short articles included in Chapter 24, we see how researchers attempt to detect such monopoles that may still exist as relics of the big bang. Grand unified theories also predict that matter is fundamentally unstable – albeit on an extremely long time-scale. It should in principle be possible for protons to decay if the theories are right, and Chapter 25 looks at experiments that have been set up to search for proton decay, and the mixed results they have had so far.

Grand unification rests on the belief that electroweak theory and quantum chromodynamics – both "quantum field" theories – are the right theories for the electromagnetic, weak and strong interactions. But gravity is left out in the cold, for there is as yet no workable quantum-field theory of gravity. An alternative approach in the quest for total unification is to turn to the symmetries underlying the outwardly diverse forces. In Chapter 26 Andrew Watson introduces the ideas of "supersymmetry". This is a scheme that attempts to make *all* particles – the quarks and leptons of matter, as well as the photons, gluons and W and Z particles that carry the forces – symmetric counterparts of each other. Supersymmetry is not a theory; more a basic structure for a theory, but it does show promise in that it offers a way in which gravity might also be included, thus bringing the four forces together at last.

Theories based on supersymmetry, like the grand unified theories, make predictions that can be tested at particle accelerators. In Chapter 27 Bill Willis describes another way in which accelerators can provide a window on creation and the more exotic regions of the Universe. He discusses the possibility of forming new states of matter in "mini-bangs" – high-energy collisions between beams of heavy nuclei. Such experiments may lead us not only to a better understanding of nuclear forces, but also of the extreme conditions that exist for instance in neutron stars. The accelerator is indeed fast becoming a tool for cosmology and astrophysics. The latest developments of this tool, and the detectors that go with it, form the subject of Part Seven. These are the instruments that can take the physicist closest to creation.

24

Grand unified theories

SHELDON GLASHOW

Theories that unify three of nature's fundamental forces cast light on the formation of matter in the early Universe. But they predict the existence of heavy magnetic monopoles and the ultimate decay of all matter.

Arthur, an intelligent alien from a distant planet, lands on Earth and spies two humans playing chess. He sets himself two problems: to determine the rules of the game, and to find a winning strategy. My metaphor describes the two great divisions of physical science. In disciplines such as atomic physics, solid state physics or chemistry, the rules are known perfectly well; the building blocks are electrons and atomic nuclei, the forces are electromagnetic, and the laws are quantum mechanics. The problem is "merely" to apply the laws to the complex issues at hand. Nonetheless, new and exciting discoveries remain to be made; it is not enough to know the rules of chess to be a grand master.

In elementary particle physics and cosmology, things are quite different. We don't know all the rules, nor even if they are knowable. The subject of "grand unification" is at this speculative frontier. It is an ambitious attempt to find fundamental unity among the diverse strands of elementary particle physics. It may be a step in the right direction, or it may be quite wrong, but the theory now being put to a decisive experimental test. For me, these are very exciting times, and I would like to share the excitement with you.

Most physicists have an unshakeable faith in the underlying simplicity of nature; it is one of our most powerful guiding principles. Time and time again, this blind faith has been vindicated. Many of nature's bewildering tricks have been explained.

What could be more different than magnetism, electricity and light? Yet, in the 19th century, James Clerk Maxwell showed that these phenomena were simply different manifestations of the same fundamental laws. He described all these, as well as radio waves, radar and radiant heat, by a unique and elegant system of equations. Maxwell's electromagnetism is the operative force controlling everything that we see, hear, smell, feel or taste.

Albert Einstein searched in vain for an ultimate simplicity wherein all the forces of nature could be treated by a single theory. He tried to unify electromagnetism with gravitation. We are still looking for such a unified theory, but we now know some of the reasons for Einstein's failure. There are simply more forces in heaven and Earth than were dreamt of in Einstein's philosophy. He was loath to consider what were once the impenetrable mysteries of the atomic nucleus. "These questions would come out in the wash," he may have felt, "if only gravitation and electro-magnetism could be put together".

The atomic nucleus is no longer such a mystery, but we have learnt that two additional forces are needed to explain its behaviour. A correct unified theory must include a description of the strong and weak nuclear forces, which are responsible for holding the nuclear constituents together and which permit the radioactive transmutation of one chemical element into another. Einstein made another critical omission. He never completely accepted the quantum theory for which he was partly responsible. "God does not play dice," said Einstein. But we are now convinced that He does.

Today's approach to the construction of a unified theory depends on two fundamental constructs: Einstein's special theory of relativity and the quantum mechanics. These are the two great revolutionary pillars of early 20th century physics. No development in physics in the past half century can be compared with these great accomplishments. The synthesis of these disciplines is called "quantum field theory", and it is within this esoteric context that all of elementary particle physics is expressed.

The first successful quantum field theory is called quantum electrodynamics, or QED, and describes the interaction between electrons and photons (or particles of light). It is not a complete theory of elementary particle physics, as it does not describe the structure of the atomic nucleus. Nonetheless, it offers an extra-ordinarily precise and concise description of certain phenomena. The magnetic properties of the electron, for example, are correctly described to 10 decimal places – the present limit of experimental accuracy. QED is generally regarded as a paradigm for a more complete theory which can include a description of the strong and weak nuclear forces.

Let me turn to the strong nuclear interactions. The atomic nucleus is known to be made up of more fundamental particles called nucleons. There are two species of nucleons: protons, which are the nuclei of the simplest atoms, those of hydrogen; and neutrons, which are unstable particles that can, however, survive when they are within a nucleus. The proton carries an electrical charge which is precisely equal and opposite to the charge of the electron. Opposite charges attract. So it is that a proton combines with an electron to form an atom of hydrogen. The neutron, like the hydrogen atom, is electrically neutral. Neutrons and protons are very small particles compared to the atoms they help to form. Once it was thought that they were truly pointlike and fundamental structures. Protons, neutrons and electrons were generally believed to be the ultimate building blocks of matter.

All this has changed. The nucleon is now known to be a composite system made up of three quarks. It is the quarks that now seem to be the ultimate constituents of nuclear matter. Two kinds of quarks, called "up" quarks and "down" quarks, are needed to construct nucleons. The electric charges of the quarks are fractions of that of the electron, normally taken as "unit charge": "up" quarks carry ⅔ of a unit of charge and "down" quarks − ⅓. A proton is a composite system containing two "up" quarks and one "down" quark; a neutron contains two "down" and an "up". Particles containing three "up" quarks or three "down" quarks are very short-lived, but they have been detected in the laboratory.

The fundamental interaction among quarks which binds them three at a time to form nucleons is known as quantum chromodynamics, or QCD. Although QCD and QED describe very different kinds of forces, there are certain profound similarities between them. Both are quantum field theories of the kind known as "gauge theories". QED leads to interactions between charged particles such as electrons and nuclei; it is QED that binds them together to form atoms. In a similar way, QCD leads to interactions which hold quarks together to form nucleons. The quantity analogous to the electrical charge of QED is called "colour" in QCD. Quarks have colour. More specifically, each "flavour" of quark (up or down) comes in three colours (red, yellow, blue, say). The nucleons, however, are "colourless". They contain one quark of each colour. When these loose remarks are translated into a precise mathematical language, we discover why it is that *three* quarks – not two or four – can bind together to form an observable particle (see Chapter 15). We also understand why it is that a free quark cannot be produced, that is, one that exists alone. A quark is coloured, and only colourless systems can exist as isolated particles. Quarks exist, but only as constituents of the particles they make up.

In the past few years, there have been a number of successful experimental tests of QCD. It now appears that we have a correct theory of the strong interactions, just as QED is a correct theory of electromagnetic interactions; though with QCD we cannot yet make predictions about nuclear processes that are accurate to 10 decimal places.

I now turn to the weak nuclear interactions – those responsible for the process of nuclear beta decay. I have mentioned that a neutron on its own is unstable. After a few minutes, it decays into three stable particles: a proton, an electron and an anti-neutrino. (Neutrinos and anti-neutrinos are massless or nearly massless particles with neither strong interactions nor electromagnetic interactions. Ghostlike, they can travel through kilometres of matter without suffering a collision.) Until very recently, there existed no theoretically satisfactory theory of weak interactions; but now there does.

Each type of force is carried by a special kind of particle. Electromagnetism is mediated by photons: the force between two charged particles can be thought to be due to the exchange of these photons. Similarly, the strong interactions are mediated by "gluons". Unlike photons, gluons cannot be seen directly because they, like the quarks, have colour. However, there is good but indirect experimental evidence for their existence. Since the 1930s, physicists have suspected that there is a special particle to mediate weak interactions: the "intermediate vector boson" or W particle.

The process of beta decay was supposed to proceed stepwise: first a neutron emits a W^- and becomes a proton; then the W^- becomes an anti-neutrino and an electron. Although this gives a useful and predictive picture of weak interactions, it is not a self-consistent picture.

In the past decade, it was realised that a sensible theory of weak interactions required the unification of weak and electromagnetic interactions. An "electroweak theory" has been developed, that has had many triumphant experimental successes. Most important, it predicted the existence of a new class of physical phenomena called "neutral currents". These have been seen, and they agree in quantitative detail with what the theory predicts. The electroweak theory, like QCD, is a gauge theory. So all three of the fundamental elementary particle forces can now be included in a self-consistent and highly predictive quantum field theory. This "standard theory" offers an apparently correct and complete description of all phenomena, excepting gravitation. Yet, from an aesthetic point of view something is missing.

QED is a theory which involves only one arbitrary parameter: the strength of the electric repulsion between two electrons. The "standard theory" involves *seventeen* such arbitrary parameters. This is clearly too large a number to be truly fundamental. Moreover, the standard theory does not explain the fact of "charge quantisation": that the electric charge of the electron is *exactly* equal and opposite to the charge of the proton.

Grand unified theories attempt to overcome these set-backs. More particularly, the theories take into account the observation that the fundamental quarks and leptons (particles *not* made of quarks) seem to come in simple "families". The first "family" consists of: the "up" quark, with charge ⅔; the "down" quark, with charge − ⅓; the electron, with charge − 1; and the neutrino, with zero charge. The first family is all that is necessary to operate "Spaceship Earth", and to fuel the Sun. However, there seem to be *three* such families of fundamental particles (see Chapter 13). The members of the other two families are the constituents of unstable particles that are produced by cosmic rays or at particle accelerators. Each family consists of two quarks and two leptons, with the same electric charge structure. We do not understand why in nature there are three families and not just one.

In grand unified theories, each of the families is regarded as a single structural unit. Particles as different as quarks and leptons are grouped together. The electrical charges of quarks and leptons hint that this is sensible: if we remember that quarks come in three colours, we see that the sum of the charges of the particles in each family is exactly zero.

I must say a word on what it means to "put together" dissimilar particles or dissimilar forces. The degree of symmetry of a system often depends upon temperature: usually there is more symmetry at high temperature than there is at low temperature. A liquid, for example, is "isotropic": all directions are equivalent. When it cools, it may form crystals. These are anything but isotropic, but possess a well-defined symmetry which is less than total isotropy. Permanent magnets offer another example. The dissymmetry reflected by the existence of separate north and south poles disappears when the magnet is heated and loses its magnetism.

Thus it is that particle physicists regard the observed dissymmetries of nature as an artifact due to the prevailing low temperature of the Universe. At higher temperatures, the innate symmetry of the Universe is revealed more clearly. Unfortunately, "high" here means very high indeed. There are three domains:

Hot   Here I mean temperatures in excess of 10^{29} K, or energies of 10^{16} GeV. Under these conditions, there is complete symmetry.

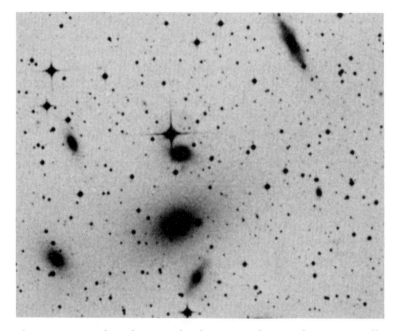

Plate 24.1   *This cluster of galaxies in the southern constellation Pavo is some 300 million light years distance from our own Milky Way galaxy, yet it is bound by the same physical laws that govern the world about us. Grand unified theories begin to draw together the seemingly different strands of cosmology and particle physics. They deal with the birth of matter, the death of matter and the fundamental nature of the Universe today. (Credit: Photolabs ROE)*

Quarks and leptons are massless and not distinguishable from one another. Strong, weak and electromagnetic interactions are evidently equivalent parts of a grand unified theory. These temperatures are today quite inaccessible. They were achieved only in the earliest moments of the big bang. Since then, the Universe has congealed, thereby losing its symmetry.

Lukewarm This is the region between 10^{29} K and 10^{15} K, or between 10^{16} GeV and 100 GeV. The lower range of energy is accessible to accelerators and at these temperatures the symmetry between weak and electromagnetic interactions has become evident with the observation of the charged Ws and their neutral partner, the Z (see Chapter 22). This is the range of energy in which the unified electroweak theory can be put to precise test.

Cool At lower energies, all the symmetry of the Universe is lost except "colour", which is at the root of strong interactions, and "charge", which generates electromagnetism. Weak interactions are observed to be weak because everyday energies are so much smaller than the electroweak scale of 100 GeV.

The simplest grand unified theory is based on the mathematical group known as SU(5). This is the unique unified theory in which the basic families of particles are simply the known quarks and leptons. There are exactly 24 particles in this theory whose exchange produces forces: the *photon*, which mediates electromagnetism, and is both massless and observable; the *three weak intermediaries*, which cause the weak interactions, and are massive particles (W^+, W^- and Z^0) which weigh about 100 GeV (in energy terms) and are also observable; the *eight gluons*, which mediate the strong "chromodynamic" interactions, and, being coloured, cannot be observed in isolation, although their indirect "footprints" are seen quite clearly; *twelve more particles*, which are very massive (about 10^{15} GeV), very shortlived, and quite unobservable. These last particles lead to a new force 10^{28} times weaker than ordinary weak interactions. This new force can convert quarks into leptons, and so can lead to the decay of protons – the most dramatic prediction of grand unified theories. All matter therefore decays, although with a very long half-life. It should take the average proton 10^{31} years to decay, give or take a factor of ten.

In a tonne of matter, the theory predicts that one proton should decay in each decade. Although this seems to be a very small effect, it can be detected. Experiments are being deployed in underground

Plate 24.2 *Digging for proton-decay in the Morton Salt Mine in Ohio, where a team from the Universities of California (Irvine) and Michigan, and the Brookhaven National Laboratory has installed its proton-decay detector of 10 000 tonnes of water. (Credit: University of California, Irvine)*

mines involving hundreds or thousands of tonnes of material: in Japan, India, Utah, Ohio, Minnesota and under the Alps (see Chapter 25).

Not only does grand unification predict the death of all matter, but it may provide an explanation for the birth of matter. Scientists have long wondered why the Universe contains matter but no anti-matter, but only today are we finding the answer. In the very young Universe, the force which produces and destroys protons was much stronger than it is today. It is believed that the same force which will lead to proton decay was once responsible for the production of matter in the first place.

Grand unification has only a few experimental tests. It predicts the value of a parameter which I invented, but which is known as the Weinberg angle. Five years ago, this prediction was in serious disagreement with experiment. With more precise experiments alone, the experimenters have changed their tune. The experimental value of the Weinberg angle is today in excellent agreement with the value that is predicted by SU(5). We shall soon see whether or not the proton decays, as SU(5) says it must. The virtues of grand unification are almost without number. It is obviously a precursor to the fulfilment of Einstein's dream of unifying *all* the interactions. It explains the fact that the photon is a massless particle. In the standard (not unified) theory, the photon could have had mass, but chooses not to. In a grand unified theory there is no such possibility. The fact of charge quantisation is likewise explained. The equality of proton charge and positron charge is forced upon us, as well as the curious fractional charges of the quarks and the zero charge of the neutrino. In SU(5), there simply is no other way. Each family of quarks and leptons is a well-defined representation of SU(5) with properties that *must be* just as they are *seen to be*. We even understand why the strong interactions are so much stronger than electrodynamics. When the Universe was young, the interactions were of the same strength. In the cold "crystalline" Universe of today, they have come apart in a calculable and well understood manner.

Perhaps the greatest virtue of grand unification is its prediction of a new class of observable phenomena. In SU(5), or in its simpler elaborations, the proton (and hence all nuclear matter) must decay with an observable lifetime. (Incidentally, we feel it "in our bones" that the proton lifetime is longer than 10^{16} years. Were it so short, we would be killed by the radioactivity produced by decaying matter within our own bodies!) This prediction comes at a very fortuitous time. We may be approaching the end of the line in our pursuit of higher and higher energies. The Large Electron-Positron (LEP) machine will cost Europe about £250 million and use a great deal of electric power (see Chapter 29). Larger machines may be built, but clearly not very much larger. Surely we can never achieve an energy wherein grand unification is apparent: such a machine would consume more power than the Sun produces. How fortunate it is that nature has provided us at least two windows into the world of high temperature: the Universe as a whole, which was baked at such temperatures during birth, and the possible decay of matter. Why is there matter? And, does it decay?

Grand unification suggests (but does not quite demand) the existence of other forms of matter that may be studied away from the high energy frontier. One such form is magnetic monopoles, particles first invented by Paul Dirac decades ago. They are to magnetism as electrons are to electricity – isolated magnetic north or south poles, unlike magnets or other known magnetic systems which bear north and south poles together. In grand unified theories such particles should exist, not as fundamental particles, but as "topological condensates" or undoable knots in the fabric of quantum fields.

The monopoles of grand unified theories are heavy, weighing about 10^{16} GeV. Their mass, converted into energy of motion, is that of a bus at high speed! Monopoles have been searched for, but never monopoles like these. So heavy are they that when placed on a table, they would fall through to the core of the Earth. Possibly, monopoles are incident on Earth in cosmic rays, perhaps as many as one monopole per square metre per day.

Grand unification may, as I have said, turn out to be merely fools' gold. But, it will soon be tested. It offers the promise of exciting discoveries in particle physics, well away from the expensive high-energy frontier. It deals with the birth of matter, the death of matter, and the fundamental nature of the Universe today. What more can we ask of a mere theory of physics?

18 September 1980

This week

An elementary particle that has escaped physicists for more than 50 years may at last have signalled its existence in an experiment at Stanford University in California. If Blas Cabrera can verify an effect he has observed only once he will be able to lay claim to having discovered the magnetic monopole, a particle that carries a single magnetic pole.

Unlike electric charges, which can occur as isolated positive or negative charges, magnetic "charges", or poles, always seem to occur in pairs: north and south. Cut a bar magnet in half and you end up with two smaller magnets, each with a north and south pole, rather than two pieces with opposite poles. This is because all the atoms in a magnetic material behave as tiny magnets, each atomic magnetic field being generated by electrons orbiting the atomic nucleus (just as an electromagnet is created by an electric current looping round a coil).

However, in 1931, Paul A. M. Dirac, now one of the "grand old men" of physics, predicted that particles carrying single magnetic poles should exist. He was trying to explain why electric charge has only certain values that are whole multiples of the basic unit of charge carried by the electron. He could explain this aspect of electromagnetism only if there also exist particles that carry similar units of magnetic charge.

Since Dirac published his predictions, physicists have searched high and low for monopoles: experiments at particle accelerators, in cosmic rays, and even in Moon rocks, but to no avail. However, they may not have been looking for the right effects.

Theorists working in the mid-1970s, particularly Gerard 't Hooft in Utrecht and Alexander Polyakov in Moscow, found that monopoles must exist if a certain class of theories are to hold. These so-called "grand unified theories" link together electromagnetism with the weak and strong nuclear forces, which are manifest only at sub-atomic dimensions. The monopoles required by grand unified theories are, however, "super-heavy", 10^{16} times the mass of the proton, or about as heavy as an amoeba!

Such a particle is so much heavier than any other elementary particle yet discovered that it could well explain why previous searches for monopoles have been unsuccessful. Cabrera's experiment, on the other hand, does not rely on any assumptions about the mass; indeed it is deceptively simple.

His apparatus consists of a loop of niobium, 5 centimetres in diameter, kept in liquid helium at a temperature only 4.2 degrees above absolute zero, or 4.2 K. Niobium is a superconductor below 9.2 K; that is, it has virtually zero electrical resistance, so once an electric current is circulating through the loop it will persist for many months, even years. However, the passage of a magnetic monopole through such a loop would cause a jump in the value of the electric current, and this is just the effect observed in February 1982.

"It is exactly the kind of signal you would expect from the hypothetical monopole," says Cabrera, but he cannot "absolutely rule out other causes". However, none of the other effects Cabrera has tested produce a signal resembling the one that could be due to a monopole.

If Cabrera has observed a monopole, the problems for physicists are far from over. Superheavy monopoles must be relics of the very earliest moments of the Universe, for only immediately after the big bang would there have been enough energy in the Universe in any one place to create such massive particles. But theory suggests that there are not enough monopoles around today to explain why one should have passed through Cabrera's small loop of niobium in such a relatively short space of time. If they are so rare, why did it take only a few months to see one? As Cabrera puts it, the signal he has observed "seems improbable whichever way you look at it". Just how improbable, he hopes to discover.

6 May 1982

Monitor

Blas Cabrera's possible detection of a monopole in 1982 aroused a great deal of excitement. But at the same time it raised the question of why he had succeeded, where the many other experimental searches had failed. One way to reconcile the results was to assume that monopoles move relatively slowly, at only 0.1 per cent the velocity of light or less. Such slow-moving particles would not be detected in conventional particle-physics experiments, which register a particle's passage by its ionisation of atoms in the detecting medium. But Cabrera's detector, a superconducting loop, would pick up a monopole passing through it irrespective of its velocity.

Astrophysical arguments also provide limits to the possible velocities of monopoles. If bound in our Galaxy they must be travelling at around 0.1 per cent the speed of light; if bound within the Solar System (as Earth is) their velocity must be nearer 0.01 per cent that of light.

Now a team from Stanford University and the University of Utah has searched for slow-moving heavy monopoles in conventional detectors in the Mayflower Mine in Utah. D. E. Groom and colleagues report finding no monopoles with velocities in the region 1.4×10^{-4} to 3×10^{-2} of the velocity of light (*Physical Review Letters*, vol. 50, p. 573). They conclude that there must be less than 5×10^{-12} monopoles per sq. cm per steradian per second in the vicinity of the Earth, and that "the Cabrera candidate is unlikely to have been a true monopole-induced event".

Interpreting data in terms of slow-moving massive monopoles is not easy, because no one is quite sure what the ionising effects of such particles might be. Sidney Drell, from Stanford University, and colleagues have calculated the effects of a monopole on the simple atoms of hydrogen and helium. They take into account, for the first time, the magnetic influence of the monopoles, and find that monopoles should deposit more energy in these materials than had been previously suspected (*Physical Review Letters*, vol. 50, p. 644). Groom and his colleagues point out that similar calculations are badly needed for other materials that are used in monopole detectors, such as argon and methane.

Another experiment, this time in the Soudan iron mine in Minnesota, has set an even lower limit on the possible number of monopoles with velocities between 10^{-3} and 10^{-2} that of light. A team from the University of Minnesota and the Argonne National Laboratory report finding no monopoles down to the limit of 4.1×10^{-13} per sq. cm per steradian per second (*Physical Review Letters*, vol. 50, p. 655).

While monopoles fail to oblige experimenters and continue to evade detection, they also fail to oblige those astrophysicists who had thought that monopoles might be the answer to the problem of the missing solar neutrinos. Neutrinos are particles produced in the thermonuclear interactions that power the Sun. The problem is that calculations of processes in the Sun's interior suggest that some three times as many neutrinos should reach Earth from the Sun as are in fact detected. One solution to the problem is that monopoles within the Sun catalyse certain nuclear fusion reactions, but not those that produce the neutrinos detected here on Earth.

J. S. Trefil and colleagues at the University of Virginia have looked into this suggestion in more detail (*Nature*, vol. 302, p. 111). Their calculations of reaction rates in the Sun's interior imply that there must be at least 10^{20} times as many monopoles in the Sun as is possible according to the limits set by the non-detection of monopoles here on Earth. They conclude that "Catalysis of fusion by magnetic monopoles appears to be another non-solution to the solar neutrino problem". It seems that monopoles continue to present more problems than they solve.

17 March 1983

Monitor

The magnetic monopole remains as elusive as ever, now evading detection by the one technique that in recent years has shown evidence for a particle carrying a single magnetic "charge", as is predicted for the monopole. In February 1982, Blas Cabrera at Stanford University, observed a signal in a superconducting coil that seemed consistent with the passage of a monopole through the coil, and which defied other explanations. Now, together with other colleagues at Stanford, Cabrera has built a new detector from three superconducting loops, set up as concentric, mutually perpendicular circles. A monopole is expected to change the magnetic flux threading the loops, and so induce a change in the circulating current. This is just what Cabrera observed in 1982, but after five months of operation the new device has shown no evidence for monopoles (*Physical Review Letters*, vol. 51, p. 1933). The latest results indicate there must be less than 3.7×10^{-11} monopoles per square centimetre per second per unit of solid angle, and that, the researchers conclude, increases the probability of a "spurious cause" for the event Cabrera observed in 1982. The search goes on.

12 January 1984

25

Waiting for the end

CHRISTINE SUTTON

Experiments can test grand unified theories by searching for the decay of protons; this quest takes researchers deep down mines and below mountains.

The ultimate fate of the Universe is a topic wide open to speculation, but in recent years particle physics and astrophysics have come almost full circle to meet each other in theories about how the Universe began and how it might end. The general trend in research in particle physics has led to a class of theories, many of which make the slightly unnerving prediction that all nucleons (the particles that form the nuclei of atoms) should ultimately decay, disintegrating into electrons and neutrinos. In other words, on a colossal time-scale of perhaps 10^{33}, or one million billion billion billion years, all matter within the Universe is radioactive! (Here one billion equals one thousand million.) Physicists have known since the 1930s that "free" neutrons – that is, those outside a nucleus – decay, each breaking up into a proton, an electron and an anti-neutrino after an average time of 15 minutes. It is only recently that theorists have come to terms with the idea that *all* protons and neutrons bound in nuclei might also disintegrate, to leave electrons and neutrinos only. In Europe and the US, as well as Japan and India, groups of physicists have put together experiments to test this bizarre notion using conventional techniques – and within an experimenter's active lifetime!

As far back as 1954 Maurice Goldhaber, together with Fred Reines and the late Clyde Cowen, pointed out that there are two basic ways to look for the decay of a nucleon. One method uses a nucleon's disappearance from a large nucleus, an event which leaves the nucleus in an unstable state so that it either breaks into two smaller nuclei (fission) or emits electrons (beta-decay) until it becomes something more stable. Experimental evidence of this sort gives the minimum limit on the lifetime of a nucleon as 10^{26} years. The second way to search for nucleon decay is to detect the products of the decay. Nucleon decays are so rare that it makes sense to look for them only in experiments in which other interactions between particles are also rare. Experiments deep underground, shielded by the rock above from the effects of cosmic rays – energetic particles from outer space – therefore make the ideal locations. Reines and colleagues at the University of California, Irvine, have increased the limit on the lifetime to 2×10^{30} years, through examining existing data from an experiment carried out jointly with groups from Case Western Reserve University and the University of Witwatersrand, Johannesburg. This experiment, set up to study neutrinos in cosmic rays, contained some 25 tonnes of detecting medium – that is about 10^{31} nucleons – and was situated 3200 m underground in a South African gold mine.

Reines emphasises that in this and other experiments the limits which emerged for the nucleon lifetime were incidental to other work. Indeed, until 1974 there was virtually no theoretical guide or incentive to go out and look for nucleon decay. Only then, according to Reines, did the theories "begin to come alive", although not sufficiently at first for the appropriate agencies to be persuaded to fund experiments. What was this theoretical revolution that has now led many experimenters to join the previously lonely trail that Reines and Goldhaber pioneered?

During 1973–74 a series of events occurred in the world of particle physics that gave credence to a new group of theories which emerged at about the same time. These "grand unified theories" regard the electromagnetic force, the weak nuclear force and the strong nuclear force as manifestations of the same underlying field – in a similar but more subtle version of the way electricity and magnetism are different visible effects of the same underlying electromagnetic field. Nature reveals the three sides to the one basic force through a mechanism known as "spontaneous symmetry breaking". You break the internal symmetry in a magnetic material, for example, when you place it in a magnetic field, and the atomic "magnets" line up in one direction. Spontaneous symmetry breaking in everyday conditions gives the electromagnetic, weak and strong forces their widely differing strengths.

There are two main contenders for the crown of "grand unified theory". One comes from Abdus Salam of Imperial College, London, and Jogesh Pati of the University of Maryland, and the other is due to Sheldon Glashow and Howard Georgi, both at Harvard University. The Georgi–Glashow version is generally more popular, partly because it incorporates the "standard" theories of the electromagnetic, weak and strong interactions which agree reasonably well with the results of experiments. Moreover, the Georgi–Glashow theory provides answers to some long-standing questions of particle physics. It tells us why quarks, the constituents of strongly-interacting particles such as protons and neutrons, should carry fractions of the electron's electric charge; why the unit of electric charge is what it is; and how many particles there must be in nature. The price physics has to pay for the solution to these and other problems is that the nucleons have a finite lifetime.

The Georgi–Glashow theory has 24 of the so-called gauge particles that "carry" the forces between particles. Spontaneous symmetry breaking divides these into several groups. There are eight "gluons" which hold quarks together through the strong nuclear force; three "intermediate vector bosons" that mediate the weak interaction; one photon that transmits electromagnetic radiation; and 12 "superheavy bosons". These last are about 10^{15} times as heavy as a proton – in other words they each have a mass of about 1.5 ng (1 ng = 0.000000001 g). Because these gauge particles are so heavy, the force they "carry" acts over extremely small distances (about 10^{-31} m) and is extremely weak. This is analogous to saying that you cannot throw heavy things very far.

Why does this lead to baryon decay? Once incorporated in the same theory, quarks and leptons (particles such as electrons and neutrinos that do not take part in strong interactions) are no longer prevented from converting one to the other. According to Georgi and Glashow this conversion occurs through the exchange of one of the superheavy gauge particles. As the force involved is so weak such interactions are very rare, and the theory predicts that the average lifetime of a nucleon should not be more than 3×10^{33} years, and could be "as little" as 10^{31} years.

Given that the theory makes predictions for the nucleon lifetime that are tantalisingly close to the limits Reines and his colleagues calculated from existing data, it begins to seem reasonable that various groups of experimenters should have begun to think about how they might try to measure the nucleon lifetime. Basically, all you need is about 10^{33} nucleons, and if the lifetime is 10^{31} years you will detect roughly 100 decays in one year – provided you can control all other interactions in the detector that might simulate the decay of a nucleon.

Not surprisingly, Reines and Goldhaber joined forces in one of the first groups to receive money from the US Department of Energy for experiments to detect nucleon decay. This project, put forward by a team from the University of California at Irvine, Brookhaven National Laboratory and the University of Michigan, centres on a roughly cubic detector containing 10 000 tonnes of water, situated 600 m below ground in a salt mine in Ohio. The experimenters chose water because it supplies about 5×10^{32} nucleons relatively cheaply and at the same time provides a medium for detecting the products of nucleon decay. On the surfaces of the cube, which has sides approximately 20 m long, looking into the volume of water, are about 2000 light-sensitive photomultiplier tubes. These detect the photons emitted through the Cherenkov effect, in which electrically charged particles

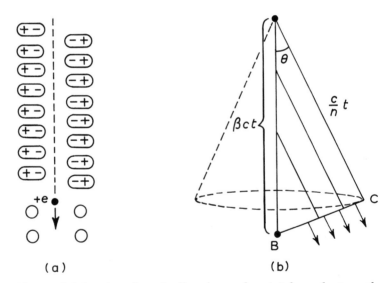

Figure 25.1 *An electrically charged particle polarises the material through which it passes (a), and atoms along this trail can radiate. If the particle moves faster than light does through the material,* then radiation from points along the path will be in phase, or "in step" (b). The radiation is emitted at an angle, θ, which is larger for faster-moving particles, and so lies along the surface of a cone.

travelling through water faster than light does *in the water* emit cones of radiation (Figure 25.1). The Georgi–Glashow theory predicts that a nucleon should most frequently decay into two particles, which move off in opposite directions to conserve momentum. The Irvine–Michigan–Brookhaven team therefore look for the tell-tale cones of Cherenkov radiation directed back-to-back. According to Georgi and Glashow a proton, for example, should decay to a positron (anti-electron) and a neutral π-meson, a good proportion of the time (Figure 25.2). The π^0 decays rapidly to gamma rays which in turn produce showers of electrons and positrons that emit Cherenkov radiation in the water.

Another experiment that has received financial backing from the US Department of Energy is the brainchild of a team from the Universities of Harvard, Purdue and Wisconsin (HPW). This group favoured a smaller detector, consisting of 1000 tonnes of water contained in a 1000-tonne vessel of concrete covered with particle detectors and assembled 600 m below ground in an old silver mine in Utah. In contrast to the apparatus conceived by Reines and his collaborators, this detector has photomultiplier tubes arranged on a three-dimensional lattice *within* the volume of the water. The phototubes that both groups will use are made by EMI in the UK, and resemble large mushroom-shaped lightbulbs with diameters of 12.5 or 20 cm. With light-sensitive linings to their almost spherical surfaces these phototubes can virtually "see" backwards – unshielded they can collect photons from three-quarters of a sphere. So with about 1000 of these "eyes" distributed throughout the volume of the water at intervals of 1 metre, the HPW detector should collect many more photons and give better information about the position of the decay, than a detector with the same number of phototubes distributed over its surfaces.

The problem that both these teams, and any others considering measurements of nucleon lifetimes, face is that of reducing the "background" from interactions that simulate nucleon decays in the detector. This is compounded by the fact that nucleons are always moving round inside a nucleus, which blurs the neat back-to-back decay pattern and garbles information about the origin of the tracks of particles. A major contribution to the background comes from neutral particles created in the walls around the detector through the interactions of particles (muons) in cosmic rays. The neutral particles could interact in the water and produce similar products to the decays of nucleons. The Harvard–Purdue–

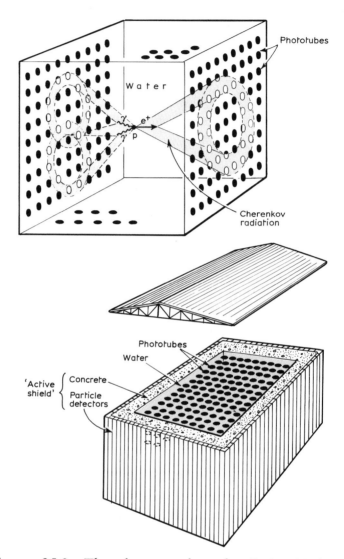

Figure 25.2 *The detector that the Irvine–Michigan–Brookhaven team has built to look for nucleon decay (top) has light-sensitive phototubes around the surface. The Harvard–Purdue–Wisconsin detector is smaller (bottom), but has phototubes distributed throughout its volume.*

Wisconsin team hopes the concrete shield and extra detectors will virtually eliminate this background, whereas the other group ignores signals from the outer 2 m layer of the water, so reducing its useful volume to 5000 tonnes. The ultimate, irreducible background comes from neutrinos that interact with nucleons in the water. Both groups estimate that 1 per cent of the time these interactions will be indistinguishable from the decays of nucleons. This would result in one indistinguishable background "event" per year per 3×10^{33} nucleons (5000 tonnes), which sets a limit on the lifetime such experiments can measure. Reines emphasises that there is a "maximum natural size" to a detector, determined by this background from neutrino interactions. With much less than 3×10^{33} nucleons a detector is too small to be useful; with more nucleons than this the background limits the lifetime the experiment can measure.

There are other groups also in pursuit of nucleon decays. Kenneth Lande heads a team from the University of Pennsylvania that has looked for nucleon decays in relatively small modules, each with a phototube, that together contain a total of 200 tonnes of water. This detector, which surrounds Ray Davis's experiment for studying solar neutrinos 1500 m below ground in the Homestake gold mine in South Dakota, was designed originally to look for bursts of neutrinos from the explosions of supernovas.

Meanwhile, the Europeans are turning to face the same challenge. A mainly Italian group from laboratories at Frascati, Milan and Turin has built a prototype detector to investigate an entirely different technique for studying nucleon decays. The project, which has the disquieting name of NUSEX (nucleon stability experiment) consists of a detector that is a sandwich of iron plates and gas-filled PVC tubes. The prototype contains 150 tonnes of iron and PVC – that is about 8×10^{31} nucleons – and is installed in an existing laboratory in the Mont Blanc Tunnel, some 3000 m below the mountain's summit. The Italians claim that their detector should be sensitive to many more types of nucleon decay than the water-based Cherenkov detectors. The team hopes to gain experience with the prototype in preparation for a full-scale 1000-tonne detector in any "second round" of experiments.

Dimitri Nanopoulos, of CERN, has said that the decay of the proton would be "the most dramatic confirmation of a grand synthesis of all elementary particle forces". But the implications of nucleon decay extend beyond the realms of the sub-atomic distances explored at the world's particle accelerators. Confirmation of nucleon decay would strengthen the claims of the Georgi–Glashow model, which, as I remarked earlier, can answer many of the "whys" of particle physics. However, it can also make important contributions to cosmology. For example, the theory says there should be three types of neutrino – no more, no less. But the "big bang" model of the evolution of the Universe also says how many kinds of neutrino there should be. What is exciting for particle physicists and cosmologists alike is that the amount of helium-4 left over from the early stages of the Universe depends on the number of neutrino types, and current estimates are consistent with there being only three different neutrinos.

The grand unified theories relate forces which in present everyday circumstances have very different strengths. Early in the Universe, however – a mere 10^{-35} s after the big bang – the temperature was high enough to correspond to the energy equivalent to the mass of the superheavy bosons. Then all interactions would have had approximately the same strength, and those that mediate nucleon decay could occur relatively easily through the exchange of superheavy bosons. Add to this picture the experimental evidence that particles can interact at different rates from their anti-particles, and the possibility that the primordial "soup" was not in a state of thermal equilibrium, and out comes the prediction that there should now be an excess of baryons over anti-baryons – a result that is entirely consistent with the lack of observational evidence for anti-matter.

If nucleon decay is discovered, then further experiments will have to show which grand unified theory is the right one. But whichever theory turns out to be correct it may have much to say about the beginning – and the end – of the Universe.

27 March 1980

Monitor

Results from an experiment in a salt mine 600 m below ground in Ohio have extended the known lifetime of the proton by a factor of 10. While it is reassuring to most people to learn that one of the basic building blocks of our bodies and the world we inhabit is more nearly permanent than we knew before, the latest data may provide problems for the physicists who are trying to establish a single theory that embraces all the four fundamental forces that shape the physical Universe as we know it.

The average lifetime predicted for the proton depends on the specific theory but most give values in the region of 10^{30} years. The only way to test these predictions in the space of a few years is to observe a sufficient number of protons and see if a few decay. The experiment in the Morton Salt Mine in Ohio is one of a number set up around the world to do just this. The apparatus in the salt mine, designed mainly by physicists from the Universities of California (Irvine) and Michigan, and the Brookhaven National Laboratory in New York, is deceptively simple. A specially-mined hole lined with plastic forms a tank for nearly 10 000 tonnes of water, or in the region of 10^{33} protons. The water is viewed by light-sensitive detectors designed to pick up the so-called Cherenkov light emitted when electrically charged particles travel faster than light does through the water. If a proton decays, its products should produce tell-tale patterns of Cherenkov light on the walls of the tank.

In a flurry of seminars both in the US and in Europe members of the team have presented the first results from the experiment. The researchers have looked for a specific decay of the proton, namely into a positron and a neutral pi-meson (π^0). After 80 days of observing they have found no proton decays of this kind. This absence of evidence translates into a lower limit to the proton lifetime of 6.5×10^{31} years. Earlier results, however, from an experiment 2300 m deep in the Kolar Gold Fields near Bangalore in India suggested that the proton *is* unstable, with a lifetime of 7.5×10^{30} years. This experiment, by scientists from the Tata Institute of Fundamental Research in Bombay, and the Universities of Osaka and Tokyo in Japan, contains 140 tonnes of iron interlaced with "proportional counters" – gas-filled detectors that are sensitive to changed particles and hence to possible products of proton decay. The researchers have found three occasions on which they claim the signal in the detectors is consistent with the decay of a proton.

A third experiment, this time in Europe, in the Mont Blanc tunnel, has also recorded one possible proton decay. This apparatus, set up by physicists from Frascati, Milan and Turin, contains 134 tonnes of iron. The result, reported in 1982, corresponds to a lifetime of about 10^{31} years, once the observing time and total number of protons are taken into account.

However, all the experiments face problems due to neutrinos:

neutral particles originating from outer space, which can travel easily through the Earth around the apparatus. These neutrinos can interact in the detector and mimic a proton decay; many physicists remain unconvinced that the "observed" proton decays are not simply due to neutrinos.

If the results from the Morton Salt Mine are correct they seem to spell trouble for one of the simplest "grand unified" theories, known as SU(5). Work presented at the 1982 high-energy physics conference in Paris gave an upper limit of 2.25×10^{31} years for a proton decaying into a positron and a π^0. If this were correct the experimenters studying the data from the salt mine should have already discovered seven proton decays.

The fact that they have seen nothing is by no means an immediate death blow for all grand unified theories. The SU(5) model is already in difficulty in dealing with the creation of particles in the early Universe. Other unification schemes suggest that other modes of proton decay should be more common, such as the decay into a muon and two π^0s. No doubt theorists are up to the challenge presented by the new results and will come back in time with new estimates, before they declare the proton stable.

3 February 1983

Monitor

Japanese physicists working on a detector 1000 m below ground in the Kamioka metal mine, 300 km west of Tokyo, have found evidence for the possible decay of a proton. Protons have long been regarded as stable, unlike neutrons and other subatomic particles which decay ultimately to protons or electrons. However, recent theories that seek to provide a unified description of nature's fundamental forces generally predict that protons should decay, albeit on a time-scale of 10^{32} years or so.

The detector in the Kamioka mine, where it lies buried to shield it from cosmic rays, contains some 3000 tonnes of water in a cylindrical steel tank. Over 1000 phototubes cover 20 per cent of the inner walls, to pick up light radiated as energetic particles traverse the water. The physicists, from the universities at Tokyo and Tsukuba, and the KEK National Laboratory, have found that on one occasion during 56 days of running the experiment, the pattern of light pulses corresponded to that expected if a proton in the water were to decay into a muon and an eta-meson, the latter itself decaying into two photons.

Other experiments searching for proton decay have found a few "events" that may be due to decays; still others have observed nothing. Only time will tell if the Japanese result is significant.

20 October 1983

26

The search for supersymmetry

ANDREW WATSON

The huge variety of our Universe often belies underlying symmetries. Now theorists are considering a "supersymmetry" that may fulfil the dream of combining quantum effects with gravity.

Consider a snowflake. Part of its beauty comes from its geometrical regularity and symmetrical appearance. In trying to make sense of the vast menagerie of "elementary" particles that have been discovered, theoretical physicists are looking for a theory with beauty like that of the snowflake. They are trying to write down equations with a high degree of mathematical symmetry. By doing so, they hope that their equations will reflect the underlying regularities of nature. The approach has already had some success in simplifying the way they describe the world, but some physicists are more ambitious: by forcing their equations to have "supersymmetry" they hope to be able to describe all the elementary particles and all their interactions – including gravity – in one mathematically beautiful theory.

If we rotate our snowflake about its centre through 60° or any multiple of 60°, the initial and final positions are indistinguishable: the snowflake has rotational symmetry. In an analogous way, the equations describing physical processes may exhibit symmetry: the initial and final forms before and after a mathematical transformation can describe the same process. In 1918, Emmy Noether of the University of Göttingen, one of this century's leading mathematicians, published a theorem relating the mathematical operation of symmetry to the real world of physics. Noether's theorem says that symmetries in the underlying mathematical description translate into quantities which physicists can observe experimentally and which are "conserved": just as the appearance of the snowflake is the same after a rotation, so the value of a

Plate 26.1 *The symmetry of the snowflake means that if you rotate one through 60° (or any multiple of 60°) its appearance remains the same. Physicists seek to find similar symmetries in the behaviour of sub-atomic particles and their forces, in order to assist them in their understanding. (Credit: HMSO Crown Copyright)*

conserved quantity does not change after a physical interaction. Conservation of electric charge, for example, is related to a mathematical symmetry of the equations describing the inter-actions of electrically charged particles; and rotational invariance is manifested in the physical world by the conservation of angular momentum.

Noether's theorem allows physicists to construct their mathe-matical theories by observing patterns among the interactions of particles in the real world. Starting from the exemplary work of James Clerk Maxwell who, in the 1860s, developed a single theory embracing the apparently disparate phenomena of electricity, magnetism and light, physicists have produced consistent working explanations of the electromagnetic force on the quantum scale also. This theory, quantum electrodynamics (QED), is perhaps the most successful physical theory devised so far.

A century after Maxwell's work, Abdus Salam, Steven Weinberg and Sheldon Glashow developed a way of intertwining electromagnetism and the weak nuclear force which is responsible for radioactive decays. Physicists regard this electroweak theory, for which the proponents jointly received the 1979 Nobel prize for physics, as a further step towards the unification of all the fundamental forces. For the strong nuclear force, which binds quarks together inside the protons and neutrons that make up atomic nuclei, physicists have developed a separate theory called quantum chromodynamics (QCD).

Each of these theories is a quantum field theory, in which various particles of matter interact by exchanging intermediary particles: these, in effect, "carry" the force from one particle to another. This exchange mechanism is one of the hallmarks of quantum field theory which has so far described three of the fundamental forces. Despite enormous efforts, physicists have so far failed to formulate a consistent theory of gravity within the framework of quantum field theory.

Physicists use quantum field theory to describe submicroscopic, fast-moving particles, since these are the participants in nature's interactions. The theory must therefore possess several special properties: it must be consistent with the quantum mechanics developed by Werner Heisenberg, Erwin Schrödinger and others in the 1920s, which describes the (slow moving) submicroscopic world; and it must be consistent with special relativity, Albert Einstein's system of mechanics for the fast-moving world. These requirements have their own symmetry conditions: for example, Noether's theorem dictates that quantum field theory represents conservation of momentum as a symmetry of the equations under translations in space, and conservation of energy as a symmetry under translations in time.

By forcing their equations to obey further symmetries, physicists hope to model the dynamics of the interactions of elementary particles. For some inexplicable reason, all the particles of which the Universe is composed are divided into two categories. Called fermions and bosons, these two categories are distinguished by the spins of the particles: protons, neutrons and electrons have spin one half (measured in natural units) and are classed as fermions; the photon, which has spin of one unit, is an example of a boson. The spin of elementary particles is a quantum mechanical effect; it is the irreducible amount of angular momentum that a particle carries with it, and is as intrinsic a characteristic of the particle as its charge or other quantum numbers.

This apparently arbitrary classification is a way of labelling particles that behave in fundamentally different ways. Fermions obey an exclusion principle, discovered by Wolfgang Pauli, which prevents two identical fermions from occupying the same quantum state. The structure of atoms is an example of the exclusion principle at work: were it not for Pauli's principle, the entire structure of electronic shells surrounding the nucleus of a chemical atom would collapse toward the nucleus. Were it not for the exclusion principle, chemistry – and hence life as we know it – would be impossible. Every fermion has an associated anti-particle, having equal mass but opposite charge and spin. Thus the anti-particle of the negatively charged electron is the positively charged positron, having exactly the same mass. When a particle and anti-particle meet, they annihilate to produce a burst of energy. Bosons are completely different. Any number of identical bosons may reside in one quantum state; there is no exclusion principle for bosons. Nor do bosons have anti-particle partners.

In modern theories, the constituents of matter are all fermions: thus the quarks which make up protons and neutrons, the protons and neutrons themselves, and the electrons that orbit atomic nuclei, all have spins of half one unit. The bosons, on the other hand, are the intermediary particles that carry the forces: thus the photon binds the electrons to the nucleus of an atom and, in the electroweak theory, fermions, such as the electron and neutrino, interact by exchanging photons, or W and Z particles, all of which have one unit of spin. The role and nature of fermions and bosons are utterly different.

Supersymmetry is a new symmetry principle that links fermions and bosons in a way consistent with the requirements of quantum field theory. For many years this development seemed impossible, but the turning point came in 1974, with the work of Julius Wess, of the University of Karlsruhe, and Bruno Zumino at CERN, Europe's centre for nuclear research, who devised a transformation which intimately linked fermions and bosons. This new symmetry transformation implies that each fermion should have a bosonic partner of identical mass, and vice versa. Noether's theorem tells us that the symmetry must correspond to a conserved quantity, linking the boson–fermion partners. For historical reasons, physicists call this the "spinorial charge". Supersymmetry thus links apparently disparate phenomena, just as 100 years ago Maxwell's theory linked electromagnetism with light.

One attraction of supersymmetry is that it suggests new ways of

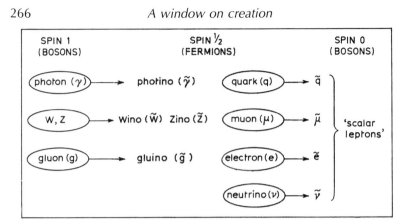

Figure 26.1 *The ringed particles are those expected from conventional particle physics. The others are their supersymmetric partners, differing by half a unit of spin.*

grouping particles together. In the past such schemes have produced predictions and enhanced understanding; can supersymmetry do the same? If we presume that a supersymmetric quantum field theory can ultimately reproduce the theories of the weak and strong forces as understood today, then supersymmetry implies the existence of partners to the "conventional" particles. Some of them are listed in Figure 26.1.

Supersymmetry has another exciting feature. If the transformation which turns a fermion into a boson is applied a second time, the result is, predictably, a fermion. Surprisingly, the final fermion is in a different position in space from the initial fermion. The supersymmetry transformation is inextricably linked with translations in space, and is the only symmetry that relates the intrinsic properties of particles, such as their spin, to their (extrinsic) position in space.

We can exploit this connection between boson–fermion symmetry and translations in space to develop a theory of gravity (see Box 26.1). Since we started in the realm of quantum physics, we would expect our theory of gravity to work in this region, the very place that conventional theories of gravity break down. The simplest example of such a "supergravity" theory was formulated in 1976 by Daniel Freedman, Peter van Nieuwenhuizen and Sergio Ferrara.

One important success of supergravity arises directly because the theory predicts a new particle, the gravitino, which is the

Box 26.1 Supersymmetry and supergravity

Symmetry transformations can be divided into two classes, global and local. In the former, the transformation is applied simultaneously to all points, represented in the diagram by an equal rotation of every point of the disc about its centre. A local symmetry transformation allows the transformation to vary from one point to the next. In our example of a disc, the circular shape is to be maintained as required by symmetry, but points on the disc surface are transformed in different ways.

Supersymmetry is a global symmetry transformation, linking particle spin properties to spatial translations. If the supersymmetry transformation is made local, then different points transform in different ways, which amounts to an acceleration between points. As Albert Einstein pointed out, acceleration is equivalent to gravitation, and thus a local supersymmetry transformation establishes a link with gravity. Theories entailing local supersymmetry are called supergravity.

There are eight different possible versions of supergravity labelled by eight conserved "spinorial charges". For more than eight spinorial charges the theory would contain particles of spin greater than two units, which are not thought to be physically relevant.

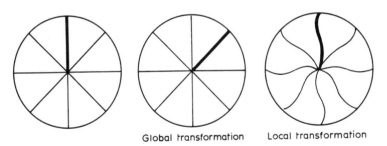

Global transformation   Local transformation

A local transformation moves the relative positions of points in space and is thus equivalent to acceleration.

upersymmetric partner to the "quantum" of gravity, the graviton. n older theories of quantum gravity, the force was transmitted only by the gravitons, and calculations invariably produced nfinities. In supergravity the new gravitino also transmits gravitaional force and, when physicists include it in their calculations, they end up with a finite result because the gravitino cancels out the infinities that plague old theories of gravity. Supersymmetry may thus be the progenitor of a sensible theory of gravity that intimately involves all fundamental particles. This prospect inevitably fuels physicists' speculations that a supersymmetric quantum field theory might be the key to unification of the four forces in nature.

It is worth clarifying the relationships of supersymmetry to the rest of particle physics. Supersymmetry is not a specific theory; it is more a condition which theories should satisfy, in a similar way that they should satisfy other symmetries such as spatial translation. Supersymmetry imposes a new condition on quantum field theory, the language in which theories of particle physics are written. What will theories of the electroweak and strong forces look like in the new supersymmetric quantum field theory? This is an open question and there are still no completely satisfactory supersymmetric theories of electroweak or strong forces.

Supersymmetry will not rise above the speculative level until physicists can link it up with the world of experimental particle physics. As yet, none of the supersymmetric partners have been observed. If supersymmetry were exact, there should be a spin-0 partner to the electron, having the same mass as the electron. No such particle has been found. To account for this, theorists have therefore suggested that supersymmetry is somehow "broken" or "hidden". We can regard this as yet another example of a recurring problem in particle physics: the problems of masses. They are never what they should be! Although such problems are no longer considered fatal to theories of particle physics, they test the ingenuity (and patience) of the proponents of those theories. More important, coping with unexpected values of particle masses, or, correspondingly, the non-observation of expected particles, introduces a degree of uncertainty that would otherwise have been absent.

Where do physicists expect to find experimental support for supersymmetry? Experiments at the Stanford Linear Accelerator Center (SLAC) in California, and the Deutsches Elektronen Synchroton (DESY) at Hamburg in West Germany, have so far identified where *not* to look. Experiments of the type performed at SLAC and DESY, where electrons and positrons are made to collide, are in some ways preferred to collisions involving protons, since the results are often less obscured by unwanted information. Experimenters at these centres might hope to witness the production of the so-called supersymmetric "scalar leptons", and $\tilde{e}$ and $\tilde{\mu}$, the bosonic counterparts of the electron and the muon. Maybe it will be necessary to wait for a larger, more powerful accelerator, such as LEP (the Large Electron–Positron collider), under construction at CERN, and due for completion in the late 1980s. It is possible that CERN's existing large accelerator, which collides protons with anti-protons, could produce another important particle, the gluino. This is the fermionic counterpart to the gluon – the spin-1 boson that mediates the strong force. However, the search is hampered because theorists cannot calculate the particle's properties exactly enough, and accelerators seem never to be quite large enough!

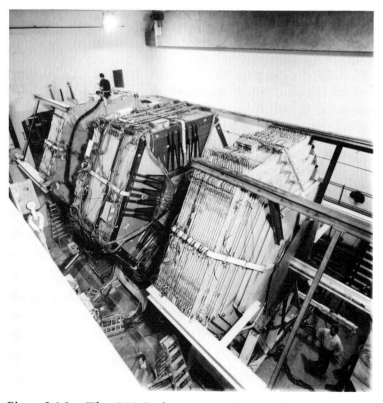

Plate 26.2 *The MAC detector studies collisions between electrons and positrons at the PEP collider at SLAC; its search for supersymmetric partners to electrons – "selectrons" – proved fruitless. (SLAC)*

Supersymmetry is not the only recent attempt at unification in particle physics. An alternative and in some ways complementary route tries to extend the techniques underlying the electroweak theory. These efforts go under the august title of grand unified theories, or GUTs to use the less noble acronym. This is a class of theories that seeks to divide the known fundamental particles into families, called generations, and unify the various forces between particles in each. Every generation contains four fermions, for example an electron with matching neutrino and two different quarks.

Grand unified theories attempt to link together the plethora of different constants that arise from the conventional view of nature. Their success has been minimal. Supersymmetry, however, does not suffer from this problem in the first place. But then supersymmetry has as yet no link with the real world, whereas GUTs use the theories of electroweak and strong forces as a starting point, completely ignoring gravity. Supersymmetry on the other hand contains an automatic link with gravity, and possesses an as-yet unrealised potential to incorporate the other forces. Both supersymmetry and GUTs will probably have a role to play. In particular, there is one problem in GUTs, known as the hierarchy problem, where two parameters have to cancel each other to an accuracy of some 24 decimal places if physicists are to produce finite results to their calculations. Supersymmetry provides a more natural solution to the problem, arranging the cancellation in a much more comely way.

15 March 1984

27

Collisions to melt the vacuum

BILL WILLIS

High-energy collisions between nuclei may provide a laboratory to test ideas on states of matter that could exist naturally only in more energetic and distant parts of the Universe.

Over the past dozen or so years, physicists have come to suspect that the physical vacuum, empty space, possesses a variety of remarkable properties that explain puzzles both in particle physics and in cosmology. We believe that early in the history of the Universe, the nature of the vacuum was distinctly different from its present state. Moreover, it seems feasible for us to flip local regions of space back into these different states by filling a large enough volume with a sufficiently high density of energy, carried by sub-atomic particles, so as to create a plasma of the quarks and gluons that appear to be the basic building blocks of matter.

Collisions of large atomic nuclei at very high energy probably satisfy the necessary conditions. In such collisions, the nuclei will be transformed from their normal composition of protons and neutrons into a soup of quarks, which will later condense back into ordinary particles. Attempts will soon be made to perform these experiments, in particular at the Super Proton Synchrotron (SPS), the largest accelerator at CERN, the European centre for research in particle physics.

In the world of classical mechanics space is the empty stage on which the events of the world take place. In combining the two major theories of modern physics – special relativity and quantum mechanics – Paul Dirac changed this view of the vacuum in the late 1920s. He found that the existence of the positron, as an "anti-particle" of the electron, with the same mass but opposite charge, was inevitable. Soon afterwards it became clear that photons, the "particles" of electromagnetic radiation, should be able to create pairs of electrons and positrons in matter.

The theory of quantum electrodynamics, which provides a precise quantum description of electromagnetism, goes still further and reveals that in empty space, pairs of electrons and positrons should be continually forming and quickly disappearing. This process does not conserve energy, but the uncertainty principle, due to Werner Heisenberg, tells us that a non-conservation (uncertainty) of energy equivalent to the mass of such an electron–positron pair will be undetectable if it lasts less than about 10^{-21} seconds.

The electron and positron formed in this way are called virtual particles, to distinguish them from real particles, which can be observed over times that are very long on the scale of the uncertainty principle. The state that has zero real particles in some particular volume is called the "physical vacuum". This coincides with the ordinary idea of an empty space, in that it contains no real particles, that is, no matter and no photons. However, as I pointed out above, according to the laws of relativistic quantum mechanics, the vacuum does inevitably contain virtual particles. We may say that the physical vacuum is as close as we can get to empty space, but that it is not quite empty.

The virtual electrons and positrons respond to an electric field in the way one would expect from their charges. So, over the short period of its existence the virtual pair tends to line up with the field, much as molecules in a block of matter tend to polarise in an electric field. For example, the strong electric field near a heavy atomic nucleus tends to produce such a "polarisation of the vacuum", and this changes the electric field felt by the electrons in the atom and thereby shifts their energy levels. This effect has been carefully measured and found to exist as predicted. The effect is small because the electromagnetic coupling is relatively weak, so that the atomic fields are little affected. However, effects of somewhat similar origin can be so strong as to lead to complete changes of the nature of the physical vacuum.

The new wave of interest in the physical vacuum stems from the notion that the approximate symmetries that we see in many laws in particle physics are not perfect only because the physical vacuum in which we live happens to have settled, perhaps randomly, into a state that lacks symmetry. This curious notion is suggested by known examples in the physics of solids, for instance. The magnetism of a sphere of iron is a good illustration. At high temperatures, iron loses its magnetism, and an isolated perfect sphere of iron will be in a state of perfect spherical symmetry. If it is cooled very slowly, until it becomes magnetic at a low temperature, it must find itself in its lowest energy state, which is one in which all the little magnets formed by the iron atoms point in the same direction. The whole sphere is then a magnet which must point in *some* direction, but the direction it chooses is entirely random.

This situation, where nature is forced to make an arbitrary choice as it cools into the lowest energy state, is called "spontaneous symmetry breaking". Another analogy is provided by imagining that a region is suddenly supplied with many automobiles. The drivers will very quickly all agree to drive on the right or left side, but the choice is arbitrary, and distant regions may not make the same choice.

The symmetries in particle physics mostly deal with quantities that are generalisations of electrical charge, used to classify the different kinds of particles. These have been assigned arbitrary names like "strangeness", "charm" and so on. We express these symmetries in geometrical language, both in order to visualise them and because the mathematics developed from geometrical problems, which describes the symmetries of crystals under rotations, provides a useful means of describing particles. However, the rotations in the symmetries of particle physics are in an abstract space, having nothing to do with the geometry of ordinary three-dimensional space. It is very handy to speak as if we were dealing with ordinary geometry, and it turns out that the use of our geometrical intuition does not lead us astray.

We believe that, like the ball of iron, the physical vacuum has picked an arbitrary direction in the space corresponding to the various kinds of charges, leading to the observed breaking of the symmetry. We can test this idea by developing a successful theory with quantitative predictions, by working out the implications for cosmology, or by experiments that modify the vacuum.

One of the most striking deviations of the real world from the simple picture of particle physics is the mysterious difficulty in observing free quarks. The quarks (and gluons) seem necessary to explain the observed families of particles, and they have been studied in detail inside protons, neutrons and mesons, using electron beams and photons. The properties of the quarks and gluons are exactly as predicted by their field theory, a generalisation of quantum electrodynamics, called quantum chromodynamics or QCD. This treats the quarks as point-like massless particles which are moving almost freely inside the proton. Surely we should be able to knock them out, and obtain single free quarks?

No such attempt has succeeded, and we now believe that the quarks and gluons carry a kind of charge dubbed "colour" which is strongly repelled by the physical vacuum in which we now live. It seems that the energy density inside a proton, due to the kinetic energy of the quarks, is large enough to flip the vacuum there into a different kind of state, which we call the "simple vacuum", in which the quarks can move freely. This little hole in the physical vacuum is called a "bag", and a mathematical model based on a simple bag has had a considerable success in explaining the spectrum of masses of the observed elementary particles.

In the early Universe, in the first few microseconds after the big bang, the energy density was high enough to keep the whole Universe in a "simple vacuum", and quarks and gluons would have moved freely except for scattering. We call this state a quark–gluon plasma. As the Universe cooled, it flipped into the state of the present physical vacuum, and the quarks would have grouped themselves into the protons and neutrons that we observe. Calculations of the physical vacuum according to the QCD theory, using techniques for which Kenneth Wilson was awarded the 1982 Nobel prize in physics, have shown just such a transition in the vacuum as the energy density is lowered.

The calculations consider a volume of vacuum inside a box, with the walls held at a fixed temperature. The energy density in the vacuum comes from different kinds of radiation in thermal equilibrium with the walls of the box. Numerical calculations on large computers predict that as the temperature is raised the colour-repelling physical vacuum should flip into the simple vacuum at a temperature of 2×10^{12} K. Temperature is just a way of describing the average energy per particle, with 12 000 K equivalent to an energy of one electronvolt, the energy an electron acquires when accelerated by 1 volt. The temperature at which the physical vacuum flips is then 180 million electronvolts (MeV) in energy units.

The vacuum absorbs a large amount of energy in making the transition, like the latent heat used in melting ice. By analogy, this transition from physical to simple vacuum is sometimes called "the melting of the vacuum". As the Universe cooled, this transition would have released a great deal of heat, and this process may have left some cosmological traces. The so-called "grand unified theories", which combine QCD with the theory of the weak nuclear interactions responsible for radioactivity, predict still other transitions. These would have occurred earlier in the expansion of the Universe and have important implications for its present structure.

An experimentalist will naturally ask if we can "melt" the physical vacuum under laboratory conditions. The temperature needed, according to the calculations, is more than 180 MeV, and this can only be generated in high-energy collisions. But how can we simulate an empty box, with the walls maintained at a well-defined temperature? It is at first hard to see any relation between that situation and a collision of high-energy particles.

We can see, though, that the essential feature is to have a region of space surrounded by walls whose function is to introduce the heat and the thermal radiation that goes with it. If the walls form a network in the space, like the walls of bubbles in a foam, so much the better. Now, if we recall that atomic nuclei are mostly empty space, with the protons and neutrons only occupying a fraction of the volume, we can see that they could be candidates for a suitable box, if we can find a way to heat them. Unfortunately, we know that hitting a nucleus with a high-energy proton usually just punches a hole in it, rather than heating up the whole nucleus. Similarly, the collisions of two protons do not have enough constituents to form a decent approximation to our box, and the constituents disperse very quickly.

When a small nucleus is struck by another large nucleus the collisions are more suitable. Although it is still true that in most collisions, most of the energy is carried away by fast particles without being dispersed throughout the nucleus, the collisions where large amounts of energy are deposited in the desired fashion are much more common. This much we may conclude from the study of a small number of collisions observed in special photographic emulsion exposed to the nuclei in cosmic rays entering the top of the atmosphere.

A collaboration of physicists from Japan and the US has flown large blocks of "nuclear emulsion" in balloons, and found an impressive event where a silicon nucleus with about 10^{14} electron-volts strikes a silver nucleus and produces over a thousand charged particles (see Plate 27.1). Measurements show that the temperature reached about 200 MeV, which is just in the desired range. This means that about 4×10^{11} eV went into "useful" thermal

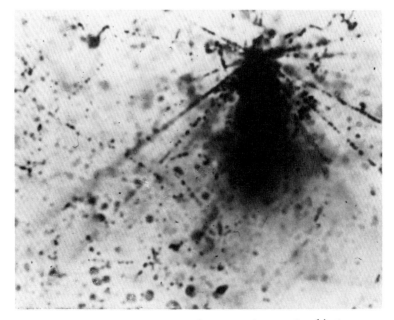

Plate 27.1 *A cosmic ray (a silicon nucleus) of 10^{14} electron-volts strikes a silver nucleus in photographic emulsion, generating more than 1000 charged particles, most of which are contained in the dark blob. (Credit: JACEE)*

energy (2000 charged particles × 200 MeV for neutrals) and the rest stayed in longitudinal motion. This efficiency is not ideal but is much better for our purpose than most proton collisions.

A system of the size produced when two heavy nuclei collide will take much longer to explode than in the case of a simple collision of two protons. The nucleons – the protons and neutrons – in the two nuclei make many collisions, and come to share the energy in a "thermal" distribution at a high temperature. They emit thermal radiation, gluons for example, and the physical vacuum in the collision volume is flipped into a simple vacuum. Quark and gluon soup fills the whole volume. The system loses heat by expanding and by radiating particles, and the space inside flips back to the physical vacuum, forcing the quarks and gluons to be confined in the ordinary particles again. This, at least, is the predicted series of events: a "little bang", followed by reconfinement mimicking the confinement of the quark–gluon plasma that occurred in the aftermath of the big bang.

The study of collisions of heavy nuclei from accelerators has been carried to energies of about 2 GeV (2 gigaelectronvolts, or 2000 MeV) per nucleon at the Lawrence Berkeley Laboratory in California. Recently the researchers there have even accelerated beams of uranium to about 1 GeV per nucleon, or about 200 GeV for the nucleus (Figure 27.1). Their studies have taught us very many facts about the collisions of nuclei. At these relatively low energies, most of the energy density comes from the compression of the nuclei in the collisions, rather than the creation of particles from the energy. Unfortunately, the best guess is that their energy is too low to observe the transition of the colliding nuclei to a quark–gluon plasma.

At CERN, helium nuclei have been introduced into the Intersecting Storage Rings, which collide two beams of particles head-on. With helium nuclei, the effective beam energy is about 1000 GeV per nucleon. With only four nucleons per nucleus, the helium collisions resemble proton–proton collisions more than those of heavy nuclei, but the researchers at CERN did observe the unexpected production of particles at large angles. Another set of experiments is planned to study further the events where large numbers of particles are produced.

CERN is now planning another big step in this area of research. The laboratory has signed an agreement with GSI, the large German nuclear physics institute near Darmstadt, whereby GSI,

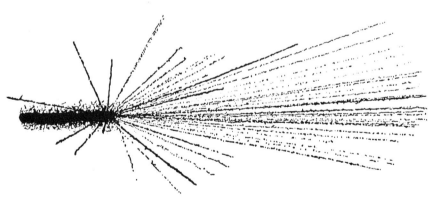

Figure 27.1 *A uranium nucleus with a total energy of 2 ×* 10^{11} *eV, accelerated at the Lawrence Berkeley Laboratory, collides in nuclear emulsion and disintegrates. (Credit: Lawrence Berkeley Laboratory)*

with participation by the Lawrence Berkeley Laboratory, will build a source for oxygen beams (with 16 nucleons per nucleus) to be accelerated in CERN's Proton Synchrotron and Super Proton Synchrotron. The energies available would range from 16×10 GeV to 16×225 GeV, with the possibility of a whole range of nuclei for targets. The first experiments might take place by the end of 1985. In addition the Nuclear Science Advisory Committee in the US has selected as the next machine to be built a high-energy collider for nuclei up to uranium. One interesting idea is to use the empty tunnel built at the Brookhaven National Laboratory in New York for the Colliding Beam Accelerator (formerly Isabelle) which has been abandoned. Plans are under way to inject nuclei as heavy as sulphur into the laboratory's existing 30 GeV synchrotron.

Sceptics ask how one can possibly learn about fundamental processes from the study of such complex and transient events, and it has not been possible to produce a plan guaranteed of success in exploring such unknown territory. The physicists, who are determined to make a try, point out that the nature of the collisions allows us to make entirely new kinds of measurements. The number of particles produced is so high that it becomes possible to measure enough quantities to give a detailed picture of the history of an individual collision, instead of obtaining merely statistical averages over many collisions. This represents an increase in power which may well compensate for the difficulties.

There are many signs that may indicate the expected transition, though most of them have not been evaluated very accurately. One of the simplest predictions deals with the distribution of the number of charged particles produced. If we consider the collision of two nuclei as the combination of N independent collisions of two nucleons, then the spread in the number of particles produced is $\sqrt{N}$ times the spread of the number of particles produced in one nucleon–nucleon collision. If all the quarks and gluons can move freely in one big "bag" (Figure 27.2), it is more likely that the spread will be simply N times as large as in a nucleon–nucleon collision. This is an increase of a factor of $\sqrt{N}$, which could be about 10, once the energy-density threshold of the transition is crossed. In other words, an increase in spread in the number of particles produced could signal the melting of the physical vacuum.

One suggestion that can be evaluated fairly well is to look at the production of electromagnetic radiation – photons and electron–

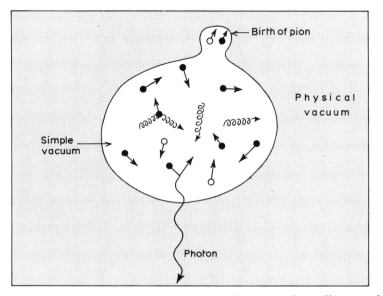

Figure 27.2 *According to present theories, the collision of two high-energy nuclei will create a big bag of "simple" vacuum, in the normal "physical" vacuum. Inside the bag, quarks (dots), anti-quarks (circles) and gluons (coils) propagate freely and scatter from each other. One of the quarks has emitted a photon, which leaves without interaction; elsewhere a quark and an anti-quark combine to form a pion.*

positron pairs – from the quark and gluon plasma. The plasma corresponds to many more moving charges than does a gas of ordinary particles, rather like many more wireless antennas. So the electromagnetic radiation from the collision should take a big jump when the transition to quark–gluon plasma is passed. The photons, electrons and positrons interact relatively weakly and so fly out of the collision volume, providing an "X-ray" of the interior, as Figure 27.2 indicates. This seems to be a feasible experiment.

The measurement of collisions producing hundreds or thousands of particles is a big challenge for high-energy physicists and their detectors. The collisions from cosmic rays were detected in nuclear photographic emulsion, and measured with microscopes scanning the tracks in three dimensions with a resolution of a thousandth of a millimetre. This technique has a great power in measuring very complex events, but it is not practical for the kinds of detailed experiments needed to identify the quark–gluon plasma.

Electronic detectors, such as the chambers that detect the tracks of ionisation left by charged particles passing through a gas, are needed to measure a high rate of events and select the rare ones of interest. In addition, detectors that can identify particular kinds of particles, are required. No one has ever measured hundreds of particles in one event in such detectors but the problem is not really unique. The particles at existing high-energy colliding-beam machines occur in clusters emitted in a very small angular cone, or "jet". The separation between the particles in the jet is very small, and this provides a difficulty for the detectors that is much the same as will be encountered in nucleus–nucleus collisions. Physicists who have studied the problem think it can be handled with the best of present techniques.

6 October 1983

PART SEVEN

Shooting sparrows in the dark

Our knowledge of the nature of matter has not come easily. The atom's secrets are locked deep within it, and only by developing the appropriate keys have physicists gained access to the subatomic world of quarks and leptons. The size and complexity of many modern experiments in particle physics often mystify the outsider. It seems paradoxical indeed that to see more detail within the atom we find we need larger and larger apparatus.

The main requirement for any experiment to probe the structure of matter is energy. This was as true for the alchemists heating substances in a retort as it is for today's particle physicists. It takes energy to break up the structures that nature has carefully forged, and to reveal the details of their contents. In atoms, the negative electrons are bound to the positive nucleus by the electromagnetic force; energy is needed to break this bond and release the electrons so that they can be observed. Likewise, the strong nuclear force binds quarks within nuclei. The nature of this force is such that single quarks are never released, but with sufficient energy we can knock clusters of quarks and anti-quarks out of nuclei, in the form of particles such as pions.

There is another, less catastrophic way in which energy can assist the particle physicist. The theory of quantum mechanics, which describes the behaviour of atoms, and electrons within atoms and so on, allows us to associate a wavelength with an energetic particle because in certain respects it behaves like a wave. (This is part of the famous "wave–particle duality", which also allows us to treat light as a stream of particles – "photons".) As with radiation, the more energetic the particle, the shorter its wavelength. And the shorter the wavelength, the more detail a wave will discern in a structure through which it passes. A toy boat, for example, will deflect small ripples on water, but will be totally swamped by large waves, oblivious to the boat's existence. So particle physicists use high energies to see detail and to prise open the secrets of sub-atomic structure. But how do they harness the energy? The answer lies in accelerating charged particles, such as electrons and protons, using electromagnetic forces. At energies that are useful, the particles are travelling virtually at the speed of light, and the machines needed to contain them can become very large, often on scales of tens of kilometres. In Chapter 28 John Lawson shows how past innovations in technology have allowed this basic technique to reach successively higher energies. The chapter also includes a "glossary" of the various types of particle accelerator that now exist. Lawson also looks to the future, and argues that another innovation is needed to reach out beyond the energies presently attainable. He introduces some of the ideas already being studied that may figure in the accelerators of the 21st century. Chapter 29, on the other hand, stays firmly in this century, and describes the accelerators that will be in operation in the late 1980s.

Once accelerated, the beams of particles are directed at "targets" – containers full of liquid hydrogen, say, or blocks of metal. Detectors around the targets trap the debris that flies out from the collisions. This may simply be the original particles deflected slightly as in a game of billiards; or the debris may contain new particles materialising from the energy available under the influence of the strong nuclear force; or it may contain particles produced in the decays of extremely short-lived states which live for only instants before transmuting to more common forms of matter. The detectors often need to be huge to trap all the fast-moving particles that fly out of the target; they need to be complex if they are to reveal precisely the identities of the huge variety of particles that can be produced.

Some detectors provide direct visual records of the passage of particles from collisions within the detector itself. The bubble chamber is perhaps the best-known example, producing trails of tiny bubbles along the tracks of charged particles. In Chapter 30 Geoffrey Hall describes how modern bubble chambers team up with other types of detector in hybrid assemblies that aim to benefit from the advantages of each of the different techniques. The chapter also includes a tribute to Luis Alvarez, who did much to make the bubble chamber one of the foremost tools of particle physics, and who received the Nobel prize in 1968.

Part Seven closes with a look at how recent advances in

technology, particularly in microelectronics, have brought into the electronics age detectors that orginated in the early part of this century with the work of Ernest Rutherford and his contemporaries. Chapter 31 also discusses how holography may help bubble chambers to keep pace with the new physics that the coming generation of acclerators will produce.

Albert Einstein referred to the art of detection in particle physics as "shooting sparrows in the dark". Today, physicists have a vast armoury of technology to help them bring a little light to that darkness. And with that faint glimmering they can perceive not only the fundamental structure of matter but reflections of the early Universe. Einstein's sparrows are perhaps more akin to birds of paradise!

28

Towards still higher energies

JOHN LAWSON

Increases in the maximum energy of particle accelerators go hand-in-hand with innovations in technology. What breakthrough will bring the next generation of particle accelerator?

Exploration of the very large and very small-scale features of our Universe requires ever more sophisticated and expensive instruments. Optical and radio telescopes have peered into space to reveal great detail about the stars and to discover remarkable and previously unknown objects. At the other end of the scale, giant particle accelerators, probing ever more deeply (by means of steadily increasing particle energy) into the constituents of the material world, have revealed a rich and complex structure that was unsuspected 40 years ago.

A study of the inhabitants of this sub-atomic domain, and the way that they interact with one another, gives us deeper insights into the laws that govern the physical world. We now have a picture in which matter is built from "elementary" particles held together by four basic forces, varying in strength and range from the strong nuclear force which holds the atomic nucleus together, to gravity which holds stars and galaxies together. The particles seem to fall into two families: the strongly interacting quarks which make up protons, neutrons and more exotic relatives, and the leptons, such as the electron and the neutrinos which do not feel the strong force. But the picture is incomplete, and the newest acclerators are presently involved in searches for the missing links. Nor do we know whether we have reached the innermost level of structure in the sub-atomic world; are quarks and leptons in fact composites of yet more elementary entities?

One way to answer such questions and to look further for the missing pieces is to build particle accelerators that can reach still higher energies. But what is evident is that building ever larger accelerators is increasingly expensive, and this is beginning to cause concern, particularly as current technology appears to be approaching its limits. What is needed, if we are to extend research in particle physics to higher energies, is some breakthrough, some new idea that can open up a whole new range in energy.

During the past 30 years there have been several "break points" where a new idea or a new technology has suddenly extended the energy that particle accelerators can reach. This progress is often represented on the "Livingston chart", named after M. Stanley

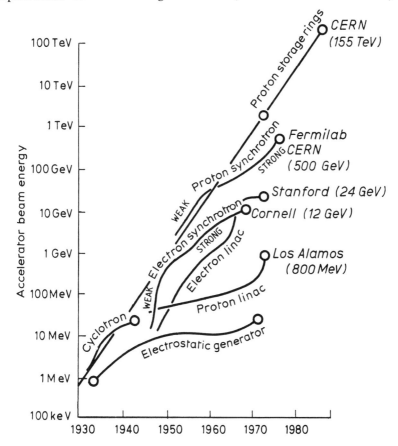

Figure 28.1 *The chart named after Livingston, pioneer of one of the first particle accelerators, shows how the energies that can be reached have increased as new ideas and technologies have improved on older ones.*

Livingston who was one of the early pioneers of particle accelerators. The chart shows how the maximum energy of the accelerated beam of particles has increased over the years, as one type of machine has superseded another (Figure 28.1). It reveals how the different accelerators have performed, but more important, indicates that over five decades the energy has increased by a factor of about 25 in each decade, or a total factor of 10 million. During the same period of 50 years, the cost per unit energy has reduced by about a million-fold, corresponding to a factor of about 16 per decade. This remarkable decrease in relative cost has arisen from the succession of new ideas. The question is, can this flow be maintained? Certainly the increase in the cost of machines cannot be; the challenge then is to find ideas that are new, but which are also economical in resources and do not demand elaborate techniques.

Before speculating about what may happen in the future it is interesting to study the past, and see how the remarkable "growth law" that the Livingston chart represents came about.

In the first decade of this century, Philippe Lenard demonstrated the transparency of matter to energetic electrons using an apparatus that was certainly a particle accelerator; but historians usually consider the subject began about 50 years ago with the electrostatic accelerator built by Sir John Cockcroft and E. T. S. Walton at the Cavendish Laboratory at Cambridge, and Ernest Lawrence's cyclotron at Berkeley in California (see Chapter 1). At the time these machines were being commissioned the mysteries of atomic structure, still baffling in Lenard's time, had become fairly well understood and the "open frontier" of physics had moved to the smaller world of the atomic nucleus. By the outbreak of war in 1939, several electrostatic generators and cyclotrons had been built, and these had provided a great deal of information on what happens when nuclei are bombarded with other particles. Researchers had observed many nuclear reactions, and the increasing accuracy of the experimental measurements had revealed a complex nuclear "spectroscopy". Theorists had put forward several semi-empirical models of the nucleus, although the detailed nature of nuclear forces was still obscure.

At the end of the Second World War, conditions were ideal for a boom in the development and construction of accelerators. The excitement surrounding the atomic bomb, and the apparent promise of cheap, clean nuclear power, ensured that nuclear physics was the glamour subject in the physical sciences. This enthusiasm, and the idealistic striving towards "new worlds" that follows periods of destructive conflict, together with obvious military interest, ensured that many people were keen to work in this new area, and that governments would be generous with resources. The war had also shown the power of scientific teamwork to tackle large projects. And techniques needed to build accelerators had been developed in other fields: the generation of power at high frequencies was required for radars, and vacuum techniques were used in the separation of uranium isotopes. Added to this, the important concept of "phase stability", invented quite independently by V. I. Veksler in the USSR and E. M. McMillan in the US, showed the way to accelerate particles to energies undreamt of before the war.

The first round of new machines, at the time considered "large", included electron synchrotrons and proton synchrocyclotrons, which could reach energies of a few hundred million electron volts (MeV), equivalent to acceleration through an electrostatic potential difference of a few hundred megavolts. Both classes of machine relied on the new principle of phase stability, in which particles travelling too fast automatically move to a position in the beam where the acclerating field is less, and vice versa for particles that are too slow. Already in the 1940s there were plans for machines to exceed 1000 MeV, or 1 GeV. These were all synchrotrons, with a magnet formed in a ring, a more economical design than the large circular magnets of the cyclotrons. During this period the technology of linear accelerators also became firmly established, providing for electrons an attractive alternative to the circular synchrotrons (see p. 292 for a glossary of terms used here).

Another big step forward, which was to inspire the next round of large accelerators, came in 1952. This was the discovery at the Brookhaven Laboratory in New York of the principle of "strong focusing". This allowed the beam of particles to be more tightly focused, so that it could travel in a smaller vacuum tube, through magnets with smaller apertures. So, with magnets of given weight and energy consumption, engineers could build a much larger ring, with correspondingly higher particle energy. Two large proton machines using this principle came into operation at the end of the 1950s, with an "orbit radius" for the particles of about 100 metres and energies around 30 GeV. The first of these was at CERN, the European particle physics laboratory near Geneva, followed very shortly by a similar machine at Brookhaven. (It turned out that, unknown to the physicists at Brookhaven, the idea of strong focusing had been patented in 1950 in Athens by a Greek engineer, Nick Cristofilos.)

In the 1950s there were many developments and many ideas tossed around, though not all of permanent value. An important centre of activity was the Midwestern Universities Research Association or "MURA" group in the US. This was inspired by the leadership of Don Kerst, inventor in 1941 of the betatron, the forerunner of the synchrotron. Although MURA failed in its bid to build a large machine, the group did help to put much of the theory on a firmer foundation, and so gave rise to valuable fundamental insights and several new concepts. Of the ideas studied by MURA, the one with the most far-reaching effect is the "intersecting beam storage ring". Beams circulating in opposite directions pass through one another at a number of points round the circumference of the magnet ring, and pairs of particles "collide". The energy released in this way, especially for particles travelling close to the speed of light, is much greater than when particles strike a stationary target.

The 1960s and 1970s were periods of steady development, though the reaction against the early enthusiasm for big spending on anything to do with nuclear energy had set in and "big" money became harder to come by. Furthermore, it now had become apparent that the higher energies were of no conceivable interest to the military. Nevertheless, the large national laboratories continued to build substantial machines, both electron and proton accelerators, and perfected the art of building storage rings.

By the early 1980s the largest machines were the giant proton synchrotrons at CERN and at Fermilab near Chicago, both with diameters of around 2 km. These reach energies of between 400 and 500 GeV. The largest electron machine is the 30-GeV linear accelerator at Stanford, California, which is 3 km long. All the most recent electron machines have been storage rings that collide beams of electrons and positrons travelling in opposite directions within the same ring. These machines reach energies of up to 23 GeV per beam. And an ingenious scheme uses the synchrotron ring at CERN to provide a proton–anti-proton collider with a total energy of 540 GeV. Work is also under way at Fermilab to convert the 500-GeV machine to a collider and to extend its top energy to 1000 GeV, making use of superconducting magnets. Perhaps the most ambitious of current enterprises is LEP, an enormous electron–positron storage ring to be built at CERN, in a vast tunnel 27 km in circumference. In its "second stage" it should exceed a total energy of 200 GeV.

So much for the past, but what of the future? All existing high-energy accelerators rely on the continuous interaction between an electrically charged particle and a guided electromagnetic wave. The particles always move more slowly than light, so the accelerating wave must do so too in order to remain in step. For such waves to exist in a vacuum they must be "tied" to a guiding structure; and if they have a significant component of electric field in the direction of propagation they can exist only at distances one wavelength or so from the structure. (Similar "evanescent", or "slow" waves, occur outside a glass block when it reflects light internally.) Thus any new idea for accelerating particles must take account of such general principles.

One popular idea for a new type of accelerator is to use laser light. There have been numerous suggestions of how this might be done, since Koichi Shimoda in Japan first came up with the idea some 20 years ago. The basic strategy rests on the enormous electric fields that large modern lasers can generate, reaching many hundreds of megavolts per centimetre. The question then is, how do we get the particles to interact with these fields? We could build a miniature linear accelerator small enough to influence the short wavelengths of laser light. A diffraction grating has the appropriate geometry, but unfortunately electrical breakdown around the structure severely limits the accelerating field such a device can sustain. Furthermore, the extremely small size limits the current, that is, the number of particles that can be accepted, and leads to almost impossible tolerances on the mechanical construction.

If we try to use a laser beam in free space, without a guide, there are two difficulties: first, the wave always moves faster than the particle, and secondly, if the wave and particle move in the same direction, the field is perpendicular to the particle's motion rather than parallel to it, so that no useful interaction takes place. We can, however, "modulate" the particle's path with a static alternating magnetic field, so that it oscillates about its main direction, picking up energy from the laser beam as it goes. This, incidentally, is the inverse mechanism of the free electron laser. Unfortunately, the electric field is still almost perpendicular to the particle's direction, and the acceleration rate at high energies is not spectacular.

If we remove the constraint that the particle must be in a vacuum, then it becomes possible to generate slow waves that do

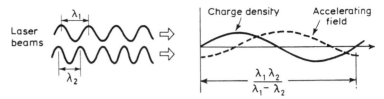

Figure 28.2 *Two superimposed laser beams of slightly different wavelength directed into a plasma cause it to separate into regions of different charge, producing a "beat wave". This has high electric fields in the direction of propagation, and travels forwards at the speed of light.*

have a component of field along the particle's path. One such possibility is the type of wave set up in a plasma by two laser beams of different frequencies. Under the influence of the laser beams the plasma separates out into regions of positive charge and negative charge in a wave that has electric fields in the same direction as the laser beams (Figure 28.2). "Back of envelope" estimates of the acceleration possible by this means look exciting, but no one has yet done a realistic analysis. Other schemes, using waves in intense electron beams, do not look so promising for very high fields.

Suppose we do not use the harmonic fields of wave motion, but rely on the high fields associatd with bunches of charged particles to "drag" other particles along. In other words, the fields created by one accelerated beam can be used to accelerate another beam of particles. Many ideas in this class of "collective" accelerators exist, the earliest being the "donkeytron" that Hannes Alfven and Olle Wernholm at Stockholm proposed 30 years ago. Fields associated with a focus in an intense electron beam are the "carrot", the particles to be accelerated the "donkey".

A more sophisticated example is the "electron ring" accelerator, foreshadowed in an unpublished note by R. Shersby-Harvie while at Malvern in 1951. It has been developed into a real concept by Soviet scientists, and studied in detail, both experimentally and theoretically, during the 1970s at the Joint Institute for Nuclear Research at Dubna and at the Lawrence Berkeley Laboratory in California. A closed ring of electrons, held together by magnetic forces created by the current round the ring, has local internal electric fields much higher than those needed to accelerate the ring

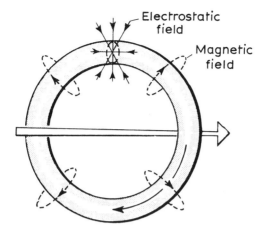

Figure 28.3 *Electrons moving in a ring generate internal electric fields much higher than that needed to accelerate the ring along its axis. The electrons generate a magnetic field as shown, which holds the ring together. A magnetic field along the axis keeps the electrons circulating.*

(Figure 28.3). The concept works, but is expensive and complicated and does not provide an accelerating field high enough to generate the high energies that we are looking for.

G. A. Voss and T. Weiland, who have been studying the "wake fields" left by electrons in the storage ring at DESY, the particle physics laboratory near Hamburg, have proposed a new idea that still has to be evaluated. Bunches of electrons that are sufficiently intense and compact leave an electromagnetic "wake" in the vacuum chamber as they move round the synchrotron ring, like a boat creates on water. It is possible that a suitable design might make these fields much higher than the fields needed to accelerate the bunches of electrons. In this way, particles following a bunch of electrons travelling very close to the speed of light might achieve very high energies.

Certainly there are many valid ways of accelerating particles. The question is, do these ideas, many of which have been around for a long time, give any promise of the breakthrough that we are looking for?

30 September 1982

Box 28.1 A glossary of particle accelerators

The basic ingredient of a particle accelerator is an electric field; this provides the force to accelerate electrically charged particles, usually protons (positive charge) or electrons (negative charge). The electric field accelerates the particles in the direction of opposite charge; thus positive particles move towards regions of negative charge and vice versa.

The simplest accelerator feeds particles from a source across an electric field (1). But the energy the particles can acquire in this way

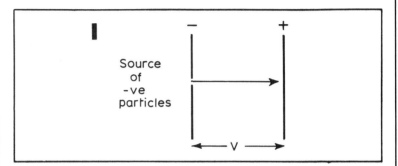

1 DC accelerator. Charged particles are attracted to the opposite electric charge and pick up energy from the electric field as they cross it. If the field is V volts and the particles have unit charge (as electrons do) they acquire an energy of V electron volts (eV).

is limited by the size of field that can be sustained; the highest practical potential is in the region of 20–30 million volts. To reach higher energies requires accelerators that give successive small "kicks" to the particles in one of two basic ways, either by using many regions of electric field or by passing the particles repeatedly through the same region of field.

In a **linear accelerator**, or linac, the particles encounter successive cycles of the alternating electric field of an electromagnetic wave; radiowaves have the appropriate wavelengths, typically in the centimetre to metre range. To accelerate protons in this way an electromagnetic standing wave is set up in a long cavity. Tubes then shield the particles during the cycles where the field is in the wrong direction (2).

Electrons, being much lighter, travel closer to the speed of light than do protons. They can thus be accelerated by a travelling

Box 28.1 cont.

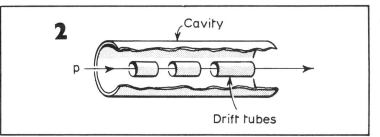

2 Proton linac. A stationary electromagnetic wave in a resonant cavity provides alternating regions of electric field; drift tubes shield protons during the cycles of negative field, so that there is net acceleration.

electromagnetic wave propagating along a hollow waveguide, provided it is loaded with some structure to slow the wave down to keep time with the electrons (3).

Another way to use a relatively small electric field in an accelerator is to make the particles traverse the same field repeatedly. A magnetic field forces a moving charged particle in a direction perpendicular both to the field and the motion; thus if the field is strong enough a particle will move round in circles. In a **cyclotron** particles released at the centre of a magnetic field region spiral outwards as they gain energy from repeated crossings of an electric field (4). A **synchrotron** matches the increase in energy by increasing the field of the electromagnets, thus keeping the particles on the same circular path until they reach maximum energy (5).

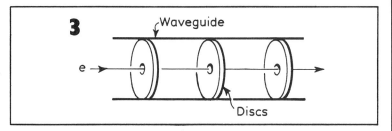

3 Electron linac. A travelling electromagnetic wave, slowed down by the discs, provides an alternating field that keeps in step with the electrons, so that the particles are always in time with the appropriate (negative) field.

Box 28.1 cont.

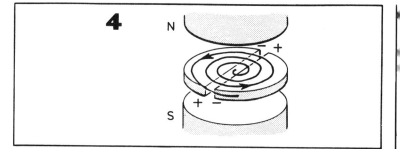

4 *Cyclotron. A charge particle released between the poles of a magnet moves in circles, while a standing wave set up in a cavity comprising two D-shapes provides an accelerating field of the correct polarity when the particles cross the Ds. As they gain energy the particles move outwards along a spiral path until they reach the edges of the magnet.*

When a particle of energy E (in GeV) collides with a stationary target, only about $\sqrt{2E}$ of this is available to create new particles; the remainder is carried away in the kinetic energy of the debris. But if two particles of energy E collide head on, then all the $2E$ is available. **Storage rings** make use of this effect by colliding two beams of particles travelling in opposite directions round rings of magnets which guide the particles (**6**). If the beams are of identical particles they must travel round two rings, interlaced to produce collisions; beams of particles and anti-particles can however travel in opposite directions in the same ring, as they are oppositely charged.

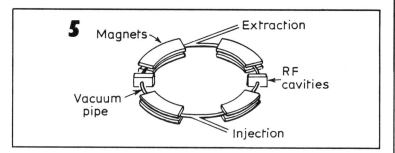

5 *Synchrotron. If the strength of the magnetic field is increased as the particles gain energy, then they can be kept on the same circular path. They pick up energy from radiofrequency waves set up in cavities interspersed between the magnets.*

Box 28.1 cont.

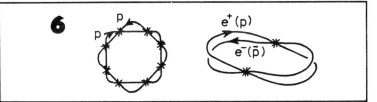

6 *Storage rings. When particles collide head-on much more energy is released than when a beam strikes a stationary target. Particles of the same charge must be stored in two rings of magnets, interlaced to produce collisions; particles of opposite charge (protons and anti-protons, for example) can be stored in two beams travelling in opposite directions around the same ring.*

29

The new breed of accelerator

CHRISTINE SUTTON

The acclerators of the late 1980s are already being built. They make the most of modern technology to reach the highest energies possible.

High energies are useful to particle physicists for two basic reasons. A beam of particles, such as electrons or protons, can probe matter in much the same way as beams of light or X-rays do. The analogy is close because the particle beam has an associated wavelength, which is smaller the higher the energy of the particles; so higher energy beams can "see" to smaller distances. Particle beams from modern accelerators probe distances as small as 10^{-16} cm, or one thousandth the diameter of a proton, which is roughly 10^{-13} cm. At this level experiments can perceive a structure to protons and neutrons, implying that they are built from smaller, more fundamental particles – the quarks.

A second reason for studying particle collisions at high energies is that the energy can convert into new forms of matter. This happens in accordance with Albert Einstein's equation, $E = mc^2$, where E is energy, m is mass and c is the velocity of light. So higher energies can also lead to new, heavier particles, as with the W and the Z which weigh in at some 80–90 times the mass of a proton.

The most common type of machine for accelerating to the energies interesting to particle physicists is the synchrotron. This uses modest electric fields to impart repeated small bursts of energy to bunches of particles as they travel a circular path through regions of accelerating field. The particles are held on their circular paths round the synchrotron by electromagnets, which provide a steadily increasing magnetic field to match the increasing energy of the particles.

In this way machines at CERN, the European laboratory near

Geneva, and Fermilab in the US, can bring protons to energies equivalent to acceleration through a potential of 500 thousand million volts – 500 times the energy embodied in the mass of a proton. And by clever adaptation, the two proton synchrotrons can reach still higher energies.

One important parameter in high-energy physics experiments is the energy available in the "centre of mass". This is a reference frame in which both beam and target approach each other with equal momentum – that is, equal velocity in the case where the two particles have the same mass, as when two protons collide. It is the total energy calculated in this reference frame that is available for conversion to matter, and the formation of new particles. When a proton of 500 gigaelectronvolts (GeV) – the maximum energy of the unmodified machines at CERN and Fermilab – strikes a stationary target, only a little over 30 GeV is available in the centre of mass. But if two 500-GeV protons collide head-on, then the collision is in effect already in the centre-of-mass frame and all the energy – 2 × 500 GeV – is available.

CERN exploited this avenue to higher energies by using its proton synchrotron to accelerate anti-protons – the anti-matter counterparts of protons with the same mass, but negative charge – in the opposite direction to the protons. The counter-rotating beams of particles are made to collide at certain points around the synchrotron, where detectors encircle the pipe carrying the beams. At 270 GeV per beam – for technical reasons the highest the synchrotron goes in this mode – the total energy is 540 GeV, which is equivalent to that available when a particle of 185 000 GeV strikes a stationary target.

By contrast, to improve its 500-GeV machine, Fermilab has exploited the technology of superconductivity – the ability of some materials to conduct an electric current with no resistance at very low temperatures. Electromagnets based on superconducting coils sustain higher magnetic fields than do conventional magnets. So a ring of superconducting magnets can contain a more energetic beam than does a normal ring of the same size. A new ring of 990 magnets, installed beneath the original magnets of Fermilab's proton synchrotron, will allow the machine to reach an energy of 1000 GeV, or 1 TeV (teraelectron volt) by 1985. The super-conducting magnets have the added advantage of using less electricity to provide the same field and therefore help the laboratory to save money.

Fermilab is also constructing a source for anti-protons and two small magnet rings to store and concentrate the beam of anti-protons prior to injection into the main accelerator. Once complete, in the summer of 1985, this project will allow the laboratory to produce proton–anti-proton collisions at a total energy of 2 TeV, thus extending the energy region that CERN has opened up and which has already proved so fruitful.

The two projects – the proton–anti-proton collider at CERN, and the superconducting ring at Fermilab – illustrate two basic means that the builders of accelerators have at their disposal to reach the higher energies that the experimenters desire. All new accelerators, either planned or being built, will incorporate either superconducting technology or the principle of colliding beams, or both. A picture of complementary work at laboratories through-out the world is beginning to emerge (see Figure 29.1) – naturally

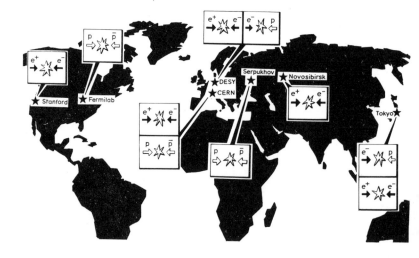

Figure 29.1   *The pattern of large accelerators across the world by the 1990s.*

tinged with some nationalistic pride and a sense of rivalry, which helps to produce the best machines possible. CERN's next big project is the Large Electron–Positron storage ring, LEP. The machine is basically a synchrotron for accelerating electrons and positrons (anti-electrons) in opposite directions; the name "storage ring" refers to the fact that once accelerated the counter-rotating bunches of particles can circulate the magnet ring many thousands of times, because only a few collisions actually occur on each encounter. In Phase 1 of LEP, which was approved by the CERN Council in October 1981, the machine will provide collisions between electrons and positrons at 50 GeV per beam. This does not sound very much in view of the energies that proton accelerators reach, but electrons pose special problems for synchrotron builders.

Electrons are much lighter than protons – a couple of thousand times so – and this in turn means that electrons radiate much more energy as they follow curved paths through magnetic fields. To make matters worse this radiated energy increases as the fourth power of the energy of the electron (or positron), so at higher energies more and more energy is lost in "synchrotron radiation", as it is known. One way round the problem is to transport the electrons on more gently curving paths, although the synchrotron ring must then have a larger circumference. This is precisely the solution adopted for LEP, which has a circumference of 27 km, taking the tunnel that houses the magnet ring under the Jura mountains, north-west of Geneva. Contrast this with the 7 km circumference of CERN's 500-GeV proton synchrotron!

LEP's gentle curves do have one advantage in that the magnets have a maximum field of only 0.13 teslas compared, for example, with 4.5 teslas for Fermilab's 1-TeV machine. The low field means that the magnets for LEP can be built to a design that costs about half that of conventional magnets. They contain laminates to shape the field correctly which are held apart by concrete. The field of 0.13 teslas is sufficient to contain electrons and positrons at 120 GeV, an energy that LEP may reach in some future phase, yet to be officially approved. To reach such an energy LEP will need superconducting technology, but in this case in the accelerating system rather than the magnets. The acceleration in a synchrotron is provided by radio frequency electromagnetic fields in resonant cavities. Cavities built from superconductor lose less energy through electric currents in their walls and so offer a less expensive, more efficient means of acceleration.

Many physicists believe that LEP represents the end of the line for electron–positron storage rings; a machine producing 350-GeV beams would need a circumference of 300 km, according to some estimates, and would cost some £5000 million – compared with about £300 million for LEP Phase 1. Burt Richter, from the Stanford Linear Accelerator Center (SLAC), in California, has argued that the cost of an electron–positron storage ring increases as the square of the energy. He has therefore become one of the main protagonists for the so-called linear colliders, for which the cost increases simply in proportion to the energy.

The idea of a linear collider is fundamentally simple: to point two linear accelerators at each other, and allow bunches of electrons and positrons to collide head on. The scheme contrasts with that of a storage ring in a number of ways. First, in a linear accelerator, the capital costs of accelerating cavities and power supplies is high, because it uses a linear chain of many cavities to reach the maximum energy, rather than using a few cavities many times as in a synchrotron. But because the particles travel in a straight line, magnets to provide bending fields are no longer necessary. A second difference is that a storage ring allows repeated collisions of the same bunches of particles; in a linear collider once the bunches have collided they are effectively lost. However, as Richter argues, at high energies the costs cross over to favour linear colliders over storage rings.

Richter's ideas have not stopped with words and calculations. To prove the feasibility of linear colliders, SLAC is working on converting its 3-km linear accelerator (linac) into a "single pass collider". Put simply, the idea is to accelerate a bunch of positrons, quickly followed by a bunch of electrons along the linac; at the end of the linac the two bunches of particles will be sent in opposite directions round two arms of a ring, where they will meet at the point diametrically opposite the linac (see Box 29.1). By improving the linac to reach 50 GeV, rather than its present maximum of 32 GeV, the physicists and engineers at SLAC hope to create collisions of the same energy as those at LEP, but two years earlier.

Even before official approval for the project, SLAC forged ahead with work on the SLC (SLAC Linear Collider), making the most of the money in its operating budget. But in 1983, President Reagan authorised the project, allocating $32 million for the fiscal year of 1984. Meanwhile SLAC had completed a "damping ring" for the electron bunches. This is a small magnet ring, 35 m in circumference, which will store bunches of electrons between pulses of the SLC and reduce them to a small enough size for acceleration in the linac. The laboratory has also built and tested prototype 50-MW klystrons (power supplies) of the kind that will be needed to boost the linac's energy to 50 GeV.

Also to be built are: a damping ring for the positrons, which will be stacked above the electron ring; a target to produce the positrons and a transfer line to take them to the beginning of the linac; the new klystrons; the two arcs of the collider, and the final

Plate 29.1   *The path of the new SLAC Linear Collider (SLC) is shown by the dotted line at the end of the 3-km (2-mile) linear accelerator. Bunches of electrons and positrons will be fed from the accelerator in opposite directions around the SLC, to collide head-on. (Credit: SLAC)*

Box 29.1 Stanford's linear collider

The electron–positron collider under construction at the Stanford Linear Accelerator Laboratory (SLAC) is based on the existing 3-km linear electron accelerator. The new idea is to use the "linac" to accelerate both electrons and positrons, and then fire them in

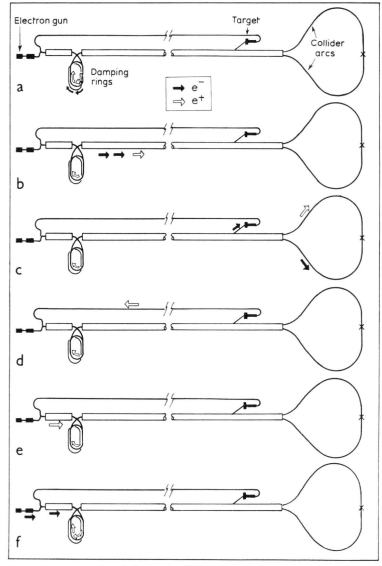

Steps in one cycle of the SLAC Linear Collider

Box 29.1 cont.

opposite directions round two arcs forming a collider ring. Electrons can be accelerated during only one half of the alternating cycle of the radio-frequency waves that the "linac" produces. During the other half, the electric field is in the wrong direction for electrons, but it is in the right direction to accelerate positrons, with their opposite charge. The SLAC Linear Collider (SLC) will use some of its electrons to produce the positrons it needs, in a complex sequence of operations that is repeated 180 times a second.

The SLC's cycle begins in the damping rings, where the lateral size of the bunches of particles is squeezed down before acceleration in the greater part of the linac (**a**). There are two rings: one with two bunches of electrons circulating at the start of the cycle, the other with two bunches of positrons.

The first step is for one bunch of positrons, followed by both bunches of electrons, to move into the linac, where they are accelerated (**b**). Two-thirds of the way along the linac the second bunch of electrons departs to collide with a target (**c**), where it produces new positrons. The remaining electrons and the positrons proceed to the end of the linac where they split off in opposite directions round the arms of the collider. Just before the collision point the bunches receive a final "squeeze" to a few micrometres across so as to optimise the number of electron–positron interactions that actually occur.

Meanwhile, the new positrons produced at the target return to the start of the linac (**d**), where they are accelerated for a short distance (**e**) before joining the bunch that has all the while being circulating the damping ring. Then two new bunches of electrons are injected from the electron source (**f**). The SLC is ready to begin its cycle again.

---

focusing system, which will squeeze down the particle bunches to about a micrometre across, so as to improve the chance of collisions. Whether all will be finished by the hoped-for date of 1 October 1986, depends crucially on how much money the project is allocated for 1985, when it needs $50–$60 million according to Richter.

Meanwhile, the Institute for Nuclear Physics at Novosibirsk in the USSR, has plans to put the full idea of a linear collider into practice. In the first phase the aim is to fire beams of electrons and positrons from two linacs at each other, with an energy of 150 GeV per beam. The total length of the collider, known as VLEPP, would be 3 km, at this stage, provided that accelerating gradients of 100 MeV/m can be created. This is much higher than gradients in the SLC, which need to be around 17 MeV/m. Tests back in 1978 showed that gradients of 150 MeV/m are possible, and last year a prototype accelerating structure 30 cm long produced a gradient equivalent to 55 MeV/m. In the second phase of the project, the aim is to extend it to reach 500 GeV per beam, for which the two linacs would total 10 km in length.

By contrast, the Soviet Union's other main research centre for particle physics, the Institute for High Energy Physics at Serpukhov, plans to build a huge proton synchrotron, called UNK. The project, which the authorities authorised in 1980, will use the existing 80-GeV proton synchrotron to feed protons into a 400-GeV accelerator with conventional magnets, which in turn will feed a 3000-GeV (3-TeV) ring of superconducting magnets. The superconducting magnets will be in the same tunnel as the normal magnets, and the two rings will be interlaced. With protons in both rings, collisions at a centre-of-mass-energy of 2.2 TeV will be possible. And with anti-protons in the superconducting ring, proton–anti-proton collisions at a total energy of 6 TeV are feasible. The tunnel for the magnets will be 19.3 km in circumference, which is small compared with LEP, but three times the size of the proton synchrotrons at CERN and Fermilab. Work is progressing in collaboration with scientists from Leningrad and Saclay in France, to develop the superconducting magnets, which will be similar to those used at Fermilab.

Building large accelerators is by no means confined to the West and the USSR. The Japanese in particular play a strong role in particle physics. At the Japanese National Laboratory for High Energy Physics (KEK), the construction of an electron–positron storage ring is well under way, having begun in October 1981. The Tristan project – Tristan stands for Transposable Ring Intersecting Storage Accelerators in Nippon (*sic*) – comprises a 3-km main ring which will collide electrons with positrons at 30 GeV per beam. Construction of the tunnel began in December 1982.

By changing to superconducting accelerating cavities, Tristan will be able to accelerate its colliding beams to 40–45 GeV, an energy for which the magnets have been designed. However, the most exciting development at KEK, will be the second stage of the project. This will see the construction of a ring of superconducting magnets in the same tunnel, which will be able to store 300-GeV protons. Once complete, Tristan will offer the prospect of electron–proton collisions at energies of over 160 GeV in the centre of mass.

Plans for another electron–proton collider, this time in Germany, have been approved by the federal government, on the grounds of "sufficiently concrete" prospects for international cooperation. Germany already has an extremely successful electron–positron collider at its laboratory, DESY, near Hamburg. The collider, called PETRA, has reached a total energy of 46 GeV. However, the proposed new machine, HERA, would involve a new tunnel, 6.45 km in circumference, buried under a nearby park! This would house two storage rings, one for electrons and one for protons. The proton ring would use superconducting magnets, again similar to those built at Fermilab, and would accelerate protons to 820 GeV; the electrons on the other hand, guided by normal magnets, would reach only 30 GeV. Both electrons and protons would be accelerated first in PETRA before being fed into the new larger rings. The project is expected to cost some DM960 millon over the seven years it will take to construct, if annual inflation is around 5 per cent.

What remains for the future? In 1978 work began at the Brookhaven National Laboratory, on Long Island, New York, on a huge proton–proton collider, called Isabelle. This was intended to collide protons at 400 GeV per beam, using superconducting magnets to steer the beams. But the development of the magnets has been beset with difficulties, and although the tunnel and experimental halls were ready in 1982, only in February 1983 were the first "production" magnets completed. Many of the original problems with the magnets apparently stemmed from the choice of a different basic design from those at Fermilab, and the design was changed significantly in 1981. But that was too late to save the project, for in July 1983 members of the High Energy Physics Advisory Panel (HEPAP) voted by a narrow majority of 10 to 7 to abandon Isabelle, now known as the Colliding Beam Accelerator.

Instead the panel chose to opt for the "Desertron", a 20-TeV proton synchrotron costing over $1000 million. The name comes from the fact that the deserts of the southern states of the US would be the only regions where land would be cheap enough to house the necessary ring of at least 150-km circumference (see next article).

Only time will tell whether HEPAP has made the right decision; whether the US will build the Desertron; and whether the Colliding Beam Accelerator is indeed "dead". Particle physics has a habit of producing the unexpected, and CERN's proton–anti-proton collider is only the second machine to find the particle it was built to discover (the first was the Bevatron at Berkeley, California, which discovered the anti-proton in 1955). Particle physicists hope to find the so-called Higgs particles which are linked to the explanation of the masses of the W and Z particles; but most of all they hope for unexpected clues as to why nature turns out to be the way it is.

11 August 1983

This week

American high-energy physicists are discussing a proposal to build a new particle accelerator that will cost more than $1000 million. It could be 200 kilometres or more across, larger than Belgium, and would be built in a remote desert region. It would certainly surpass other giant machines planned by the European Organization for Nuclear Research at Geneva (CERN) and the Deutsches Elektronen Synchrotron at Hamburg.

American physicists have woken up to the success of European accelerators – symbolised by the discovery of the W and Z particles at CERN – and are desperate to regain lost ground. Speaking at a meeting of the American Association for the Advancement of Science in Detroit, Robert R. Wilson, the former director of the Fermi National Accelerator Laboratory in Illinois, described proposals for the ultra-high energy accelerator. The machine would produce two proton beams that would collide head-on, giving interactions with effective energies of 40 million million electron-volts.

The new machine would be built in a near circular ring. To reach the enormous energies that the physicists want it would have to be huge – estimates of the circumference vary between 80 km and 240 km. The actual size would depend on the type of magnets used to keep the particles moving on their circular path.

Among the possible sites for such a ring are the region south of Albuquerque in New Mexico, near the Kitt Peak observatory in Arizona, the salt deserts in Utah, and in Texas, somewhere between Dallas, Austin and San Antonio. The barren remoteness of these locations has earned the project the nickname, the Desertron. An alternative would be to build the accelerator in an

Figure 29.2 *Possible sites for the Desertron.*

underground tunnel, possibly near Fermilab in Illinois or the Brookhaven National Laboratory in New York State.

The cost of the machine will depend on the type of technology chosen and the size and site of the machine. Using superconducting magnets providing a 5-tesla field, the cost would be around $2.7 thousand million for a 125-km ring. With more powerful magnets – say 10 tesla – the ring could be smaller, possibly 80 km across. This option might bring the cost down to $2 billion. Another idea is to use conventional iron magnets, which could provide a maximum field of 3 tesla. This would make the machine much bigger, perhaps 240 km across.

9 June 1983

Technology

When the world's first large superconducting proton accelerator broke the record for high energies in July 1983, superlatives became the order of the day. The machine, at the Fermi National

Accelerator Laboratory (Fermilab), near Chicago, involved some remarkable feats of engineering and technology, and marks the fulfilment of a dream that began over 10 years ago. In the process, the laboratory became the world's biggest user of superconducting materials, set up the longest helium transfer line and built the largest helium liquefier.

The new accelerator is a synchrotron, based on 990 superconducting magnets. The magnets steer and focus a beam of protons on repeated trips through accelerating stations in a circular tunnel, 6.4 km in circumference. After many thousands of circuits, the protons reach energies of several hundred thousand million electron volts each – roughly the energy in sunlight falling on a square centimetre in a second. The new magnets nestle beneath Fermilab's ring of conventional electromagnets which have steered protons to some of the world's highest energies over the past 10 years. This original synchrotron now acts as "injector" to the more powerful superconducting machine.

Superconducting magnets are electromagnets in which the current-carrying coils are made from material that can conduct a current without resistance, provided the temperature is low enough, generally only a few degrees above absolute zero (0 K or $-273°C$). The lack of resistance means that a superconducting magnet can produce the same magnetic field as a conventional electromagnet for much less power. In addition, superconducting magnets can sustain higher magnetic fields than can normal magnets, in which the maximum field possible is about 2 teslas. (The Earth's field is roughly one ten-thousandth of a tesla.)

These qualities make superconducting technology an obvious choice for particle accelerators, where power consumption is high – Fermilab's conventional machine consumes 90 MW when operating at 400 GeV (1 GeV = 10^9 eV). Moreover, higher magnetic fields provide a greater bending force, allowing the protons to be accelerated to higher energies in the same size of ring.

When Fermilab built its first synchrotron, the technology of superconducting magnets was still very uncertain, so the designers opted for normal magnets. However, they were far-sighted enough to leave room in the tunnel for a future generation of superconducting magnets.

The new accelerator has 774 dipole magnets to steer the beam on its circular path, and 216 quadrupole magnets for focusing. The superconducting wire is of a design known as Rutherford

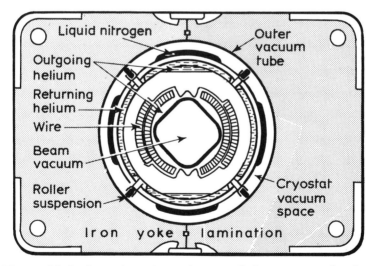

Figure 29.3 *A section through one of the superconducting magnets of Fermilab's newly equipped main ring, just visible in the aerial view in the distance.*

cable, developed at Britain's Rutherford Appleton Laboratory. One cable consists of 23 strands of copper wire around 6 mm thick, in each of which is embedded more than 2000 filaments of the superconducting alloy niobium–titanium. The magnets at Fermilab required the manufacture of some 50 tonnes of this wire, which would stretch to nearly 30 000 km, almost girdling the Earth.

The coils of each magnet sit around the pipe along which the protons travel, and which must be kept at very high vacuum to avoid collisions with stray molecules of air. The coils themselves fit inside their own cylindrical cryostat – or "vacuum flask" – cooled to 4.5 K by a flow of liquid helium. Warmer, boiling helium returns to the nearest refrigerator for recooling through an outer concentric vessel. Liquid nitrogen keeps the whole assembly cool. Refrigerators at 24 service stations around the ring take the helium returning from a chain of magnets, cool it and pump it back.

A central liquefying station, built with motors and compressors from a surplus Air Force oxygen liquefier, supplements the supply. With a capacity for producing 4000 litres of liquid helium an hour, it is the largest plant of its kind in the world. A transfer tube over the main ring connects the central liquefier with the 24 service stations. It has 24 expansion boxes, where special flexible pipe can cope with 16 m of contraction when the system is cooled.

In the superconducting synchrotron's first attempt at accelerating, it reached an energy of 512 GeV, breaking the previous world record of 500 GeV shared by Fermilab and CERN, Europe's centre for particle physics. However, its magnets can sustain fields of 4.5 teslas, sufficient to control protons up to 1000 GeV, or 1 TeV (teraelectronvolts), thus giving the project the name of the Tevatron.

In the coming months the physicists and engineers at Fermilab will be learning how to extract the proton beam from the machine safely, without depositing energy in the superconducting magnets. This could prove a disastrous mistake, because a small increase in energy can make the superconductor revert to a "normal" state, in which case it must suddenly release the energy stored in the magnetic field – a process scientists call quenching. The dipoles in the new ring store about half a million joules, sufficient energy to heat a kettle-full of water to boiling point, which is released almost instantaneously. Much of the research effort that has made this project possible has gone into designing circuits to sense a quench in a magnet and extract the energy safely.

Engineering at low temperatures brings a host of unusual problems. The dipole magnets, each over 6 m long, shrink by nearly 2 cm on cooling to the operating temperature of 4.5 K. It is crucial to the synchrotron's operation that the magnet coils remain in position as the system cools, and that the magnets remain perfectly aligned around the ring. Indeed, the physicists and engineers at Fermilab have had surprisingly few difficulties to overcome in accelerating a proton beam around the new machine, which reached successful operation more quickly than many conventional synchrotrons.

Computers help to monitor and control every aspect of the accelerator's operation. Colour displays reveal the proton beam's position to within 1 mm round the ring's 6.4-km circumference. Automatic computer control looks after the complete refrigeration system, involving some 350 feedback control loops. Leon Lederman, Fermilab's director, says that the accelerator's success has made him "ecstatic about the use of computers".

But there is another important lesson to be learned. Superconductivity has seemed full of technological promise for many years. Now Fermilab has proved that its use on a large scale is possible.

21 July 1983

30

The ubiquitous bubble chamber

GEOFFREY HALL

Bubble chambers flourished in the late 1950s and 1960s with the newly developed accelerators, but later were outpaced by electronic detectors. The 1980s have seen the advent of the hybrid chamber, which makes the most of both technologies.

Experiments in high-energy physics in the past have fallen into two categories: bubble-chamber experiments, where the liquid of the chamber provides both the target and a means to identify the particles produced in an interaction; and "counter" experiments in which large and complicated arrays of electronic detectors are used to locate and identify the collision products. More recently distinctions have blurred as the technique of the hybrid bubble chamber has emerged, employing methods of both kinds.

In 1952 Donald Glaser observed bubbles form along the tracks of cosmic-ray muons in a glass capsule of liquid ether; this was the first bubble chamber. Since then the technique has advanced considerably in technology and bubble chambers have produced millions of pictures of high-energy particle interactions. The late 1950s and 1960s was a period of exciting success when many new short-lived states were discovered and investigated, and the symmetry schemes proposed to explain particle properties were tested and began to be confirmed. However, it became obvious that other detectors can not only collect data more rapidly but can also do many experiments which were previously the province of the bubble chamber. Obsolescence became a real possibility and a large effort was made to use bubble chambers in experiments for which they are especially suited and to improve their capability to gather data at a high rate.

The principle of the bubble chamber is that a volume of liquid (usually hydrogen but sometimes a heavier material) is maintained under pressure at a temperature, above its boiling point, at which it would normally vaporise. In the case of hydrogen this is about −250°C. Particles like π- of K-mesons are fired into the chamber where they interact with atomic nuclei. As electrically charged particles traverse the liquid they deposit a minute fraction of their energy by ionisation along their path and, if the pressure is then released, local boiling results with small bubbles being formed. When the bubbles have grown to a suitable size (about half a millimetre in diameter) the chamber is momentarily illuminated and photographs taken from several directions so that a three-dimensional reconstruction of any interaction (event) can be made. It is normal for the chamber to be placed in a strong magnetic field so charged particles follow curved trajectories and their momenta as well as directions can be extracted by measuring the photographs and analysing the results with the aid of computers. The bubble density along a track is related to the ionisation and this can distinguish between different types of particle because, for energies up to about 1.5 GeV, the amount of ionisation produced by a particle with a given momentum depends strongly on its mass.

A typical experiment requires pulsing the chamber (by alternately increasing and releasing the pressure) and photographing every expansion until several hundred thousand pictures have been taken. Then the film has to be examined frame by frame, a process called scanning, to find events which are afterwards accurately measured, usually by automatic machines specially designed for this purpose. When the momenta and angles of the tracks have been computed the search for interesting features of the reactions can begin.

The principle advantages of this technique are the unique ability to see every charged particle involved in the collision; and the possibility of identifying by ionisation and momentum the individual particle types. The detection of neutral particles requires either the observation of a decay into charged products, e.g. $K^0 \rightarrow \pi^+ \pi^-$, or for inferences to be drawn from the application of conservation laws such as the conservation of energy and momentum. Electric charge itself is believed to be an absolutely conserved quantity so it is possible to be sure that all the charged products of an interaction have been observed by comparing the sum of the charges before the collision with that after it.

Very short-lived particle decays can be observed with high efficiency. One particular triumph was the observation of the Ω^-

(see Chapter 3). This important discovery confirmed theoretical ideas relating hadrons, the family of particles which feel the strong nuclear force, and laid foundations for the quark model. Quarks themselves are the supposed truly elementary constituents of all hadrons although they have not been observed directly.

On the other hand, counter experiments study reactions by means of several individual detectors, each with a specific purpose. Many ingenious devices exist, some of the more commonly used of which have been used with bubble chambers and are described below (see also Chapter 31).

Arrays of multiwire proportional chambers are often used to pinpoint the charged particles crossing them by multiplying and collecting the electrons produced by ionisation of the chamber gas. The individual, closely spaced, wires act as positive anodes to which the negatively charged electrons migrate and a small current is measured in the wire closest to the original track.

Čerenkov detectors are used to collect the light produced by a charged particle travelling at a speed greater than the velocity of light in the Čerenkov medium. In conjunction with momentum information the type of particle may be identified from its mass as either π, K or proton.

Plastic scintillation counters, as the name suggests, convert the energy deposited by the charged particle into light which is collected at a photomultiplier tube and transformed into an electronic pulse. The counter can be used to provide coarse spatial, or precise time, information.

As well as these detectors a counter experiment has a target of suitable material (again often hydrogen because of its essential simplicity) and magnets to analyse the interaction products by deflection, the amount of bend being proportional to momentum.

Of course the great advantage that an electronic experiment has over the bubble chamber is the speed at which data can be accumulated. Even though the apparatus has associated with it a certain "dead time" (the period in which the equipment is unable to function as a detector because it is recovering from the passage of particles through it), this is rarely longer than a few hundred nanoseconds. Compare this with a bubble chamber which, when pulsing once per second, has an effective dead time of almost an entire second. Modern accelerators can supply enormous numbers of particles each second so it is in the experimenter's interest to use these beams as effectively as possible, especially when one recalls the enormous cost of building and operating such machines.

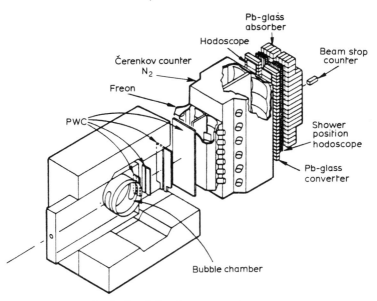

Figure 30.1 *The hybrid facility at SLAC comprises a variety of particle detectors in addition to the bubble chamber. The Čerenkov counter distinguishes pions, kaons and protons emerging from the interaction; the lead(Pb)–glass reveals photons from the decay of neutral pions, for example.*

With such thoughts in mind, bubble-chamber physicists considered how their apparatus could be more effectively used to acquire high-energy data. Many experimenters began to use neutrinos in very large chambers where many tonnes of liquid are necessary to have a reasonable chance of seeing an interaction. Neutrinos, like quarks, are thought to be elementary and are elusive because they interact with other matter only via the weak nuclear force, in contrast to hadrons which experience the strong nuclear force and electromagnetic force too. (Gravitational effects are too small to worry about.) Being without charge, neutrinos penetrate right inside the nucleus and probe the individual quarks. In these difficult experiments the slowness of the bubble chamber is not necessarily a disadvantage as, even with the most intense beams, interactions still occur infrequently. The complexity of many of the collisions, where dozens of secondary tracks are sometimes produced, also lends itself to the bubble-chamber technique. Many important results have been established, perhaps most notably the discovery of weak neutral-current phenomena where, for example, a neutrino scatters elastically from a nucleus (see Chapter 21).

Another approach was to expand the chamber more frequently. New chambers have been built to cycle considerably faster than once a second. And great success has been achieved with an existing chamber at the Stanford Linear Accelerator Center (SLAC) in California. There the one-metre-diameter bubble chamber, originally designed to run at a frequency of 1 Hz, is now running comfortably at 15 Hz – and there are expectations that even this will be exceeded, a considerable feat.

This development alone, however, is not sufficient to ensure the competitiveness of the technique since enormous quantities of film would soon overstrain the analysis capability of even the largest experimental group. It was quickly realised at SLAC that a good way to take advantage of the increased capability of the rapid-cycling bubble chamber is to operate it selectively and not simply to take a picture on every expansion. The problem is to decide when an interaction of greater interest than usual has taken place. To this end the chamber has been equipped with electronic detectors, mainly in the path of the outgoing particles, which have the dual purpose of providing a "trigger", or signal to the cameras to take a picture, and of improving the resolution of the apparatus.

The bubble chamber with its ancillary electronic detectors is usually referred to as a "hybrid system" and there are several examples in operation. At SLAC the hybrid facility comprises the chamber plus five multiwire proportional chambers (PWCs), a large Čerenkov counter, which distinguishes πs, Ks and protons emerging from an interaction and a Čerenkov in the path of the beam to recognise the type of beam particle (Figure 30.1). Additional scintillation counters allow more stringent requirements to be made of the trigger. The PWCs give rise to the improved resolution by effectively extending the lengths of the tracks which are measured, an important feature for high-momentum particles. Experiments have been performed using beams with energies up to about 20 GeV.

The hybrid technique is well matched to the electron accelerator at SLAC which provides short bursts of particles many times a second. At Fermilab, near Chicago, the giant proton synchrotron sends out its several-hundred GeV beams in much longer bursts every few seconds. At very high energies interactions occur less frequently and the hybrid system there has been operated with a different philosophy. A picture is taken on every expansion and the electronic equipment stores accurate information which is used later in the analysis stage.

Neutrino experiments have also been enhanced with hybrid techniques. The large bubble chambers at CERN and Fermilab use external particle identifiers particularly to distinguish muons from hadrons. Large amounts of steel or concrete are provided in which the hadrons dissipate most of their energy in collisions, while the muons penetrate to wire chambers embedded in the absorber.

What defines a useful trigger for a hybrid experiment? It is essential to have a particle which penetrates the downstream apparatus so that a signal to take a picture may be generated. Then, to use the facility to its fullest extent, one chooses to investigate reactions in which it is advantageous to see all the tracks in the collision. Some reactions studied at the SLAC hybrid facility have been: $\pi^+ p \rightarrow K^+ Y^*$ and $K^- p \rightarrow \pi^- Y^*$. (The Y^* is a hyperon, or heavy proton, which carries the quantum of strangeness, a quality rather analogous to electric charge.) These two reactions are closely related and provide some interesting tests of the quark model of hadrons. In both, strangeness is exchanged between the colliding particles, and the produced Y^* decays rapidly in to a Λ^0 (another hyperon) and a π-meson. Usually the Λ^0 also decays within the visible volume (into $p + \pi^-$) and can be recognised by its characteristic "V" in the chamber. The fact that this decay can be observed means that the quark model predictions can be tested more thoroughly.

The trigger in these reactions is provided by the outgoing meson (either π or K) which usually emerges with high momentum from the bubble chamber and can be identified in the large Čerenkov counter. But, before a picture is taken, the computer which handles the reading in and storage on magnetic tape of the data from the counters calls into action a short specialised program. The program calculates, in the 3 ms available before the bubbles in the chamber have grown large enough to be photographed, the position of the interaction and the momentum of the out-going meson. Since some collisions occur in the walls of the chamber, and in regions of the liquid not clearly visible to the cameras, a large saving of pictures and scanning time can be made by recognising them. And by requiring the momentum of the meson to be large the most interesting reactions are selected.

Naturally some sacrifice has to be made for the opportunity of acquiring rarer data and here this comes in the extra attention to

Fig. 30.1   *The Big European Bubble Chamber (BEBC) is 3.7 m in diameter and forms a large target for neutrino interactions. The structures being installed around the outside of the chamber form the external muon identifier, which is especially useful when picking out neutrino interactions. (Credit: CERN)*

detail required to ensure that the rejection of unwanted events is high and the acceptance of interesting ones is both large and well understood. A big part is played here by the simulation of interactions using computer programs which follow particles through the system to determine their behaviour. Many of these techniques have been developed by counter physicists though it is new to apply them to bubble-chamber experiments. A good deal of this work goes on in the planning stages of the experiment so that, by the time the data are finally analysed, the experimenters hope it will be a simple matter to apply the appropriate corrections.

Many of the developments at SLAC and Fermilab supplied groundwork for an ambitious project at CERN. This is the European Hybrid Spectrometer, a large and complicated hybrid system incorporating a brand new rapid-cycling bubble chamber, built at Britain's Rutherford Appleton Laboratory, and several other new devices. (Further developments in bubble chambers are also included in the next chapter.) Many questions of high-energy physics remain unanswered and experimenters have approached them in many ways. But if hadrons are made of quarks, as most physicists believe, then almost the only way of studying the interactions between quarks is to bombard one hadron with another. The bubble chamber still seems to have a future for doing that for some time to come.

20 July 1978 (p. 189)

Nobel prizes

The 1968 prize for physics, goes to Professor Luis Walter Alvarez of the Lawrence Radiation Laboratory (LRL) at the Berkeley campus of California University. The citation for the award to this 57-year-old physicist states that it is for his contributions in the field of elementary particle physics. In particular, Alvarez has been the leader and driving force behind a highly productive research group at Berkeley. Working with the big 30-GeV proton accelerator at Brookhaven — the twin, to all intents and purposes, of the 28-GeV machine at CERN, Geneva — in the 1960s they have discovered many new particle resonances.

These phenomena, which did not appear until the advent of powerful proton accelerators, are short-lived states of matter scarcely substantial enough themselves to justify the term "elementary particles". They are, nevertheless, a highly important development in the study of the fundamental nature of matter, for they tell us that matter can be further subdivided in certain ways which fit a pattern.

But Alvarez is more than the scientific impresario leading a prominent research group; as a man who is essentially an experimental physicist he is directly responsible for many of the sophisticated developments in instrumentation and techniques that have led up to the present streamlined procedures for studying particle interactions.

In 1946–47 he built, at Berkeley, the first proton linear accelerator embodying modern principles. In his 32-MeV machine he first tried out the idea of focusing magnetically a proton beam which was being accelerated by an electromagnetic wave along a cylindrical waveguide. The magnetic focusing successfully controlled the undesirable tendency of the proton beam to diverge as it travelled along the pipes.

Although proton linear accelerators are capable of reaching only comparatively low particle energies, their importance lies in the fact that they are used as the preliminary injection systems for much more powerful synchrotrons operating at tens and hundreds of GeV. They are thus an essential part of high-energy physics research.

Linear accelerators also have a separate role as heavy-ion accelerators used in atomic physics. At LRL, for instance, the HILAC machine, in which elements 102 and 103 were created, accelerates heavy nuclei like carbon, argon and neon.

After the machines come the detection systems for looking at particle interactions. One of Alvarez's major contributions has been the development of the hydrogen bubble chamber from the highly ingenious, but small, brainchild of Donald Glaser – the Berkeley professor who received the Nobel prize for its invention in 1960 – into the modern giant cornucopias which pour forth their vast numbers of particle-track photographs today.

The bubble chamber, a direct descendant of C. T. R. Wilson's famous cloud chamber, contains a liquid in a superheated condition. The passage of a charged particle through it ionises molecules within the liquid and sets up very local boiling, producing a string of tiny bubbles to mark the track of the particle.

Just as in a cloud chamber (where the gas is supersaturated and condenses as droplets along the particle track) dust may cause precipitation, so in a bubble chamber similar impurities may stimulate unwanted boiling. After Glaser had invented his prototype bubble chamber there was considerable controversy as to whether or not bigger ones could be made successfully. Larger chambers would necessitate using combinations of different materials in their construction, and some physicists argued that the superheated hydrogen would boil in an undesirable fashion at the junctions between dissimilar materials.

Alvarez believed that this difficulty could be circumvented if the release of pressure, introduced to make the liquid superheated, could be achieved fast enough. His hunch proved right, and opened the way to the present generation of large, so-called "dirty" bubble chambers – one of the particle physicists' most valuable tools for probing what happens when fast particles collide with targets. Glaser's invention dates from 1953; only three years later Professor Alvarez was producing excellent results with the 180-cm (72-inch) bubble chamber he built at Berkeley. The much larger volume which it became possible to examine for particle interactions as a result of this development led ultimately to an enormous data-handling problem. Here again, having played such a large part in creating the situation, it was Alvarez who came to the rescue of researchers inundated with the million or so photographs which every large bubble chamber turns out each year.

He and his colleagues developed many of the basic techniques for automatic viewing of the pictures, scanning the particle tracks, and measuring their momentum from the nature and curvature of the tracks; their first system revelled in the name of "Franckenstein". More recently Alvarez has been responsible for developing scanners which read the bubble chamber photographs in a spiral fashion.

All of these experimental advances have, in one way or another, greatly increased the chances of being able to spot and manipulate the rare, short-lived particle resonance states which in recent years have become the main preoccupation of high-energy physicists.

7 November 1968

31

Chips, charm and holograms

CHRISTINE SUTTON

As accelerators reach higher energies and probe smaller distances, experimenters call upon an ever-increasing armoury of techniques to keep track of the particles.

In the late 1980s a new generation of particle accelerators will begin life. These "atom smashers" will probe the nature of matter at higher energies and in more detail than ever before. A complete battery of detectors to capture and record the high-energy collisions of particles at one of these machines will weigh many tonnes and cost in the region of £20 million. The detectors will epitomise the art of harnessing new technology to trap elusive fragments of matter. They will also highlight an old paradox: why so big, to see something so small?

To delve deeply into the structure of matter requires energy: energy to disrupt the forces that hold the basic constituents of matter tightly within the atom and its nucleus; but also energy to see more clearly, in the following way. We can use a beam of high-energy particles to probe matter in the same way that we use beams of light or X-rays. The detail we discover with our probe depends on its wavelength; quantum theory allows us to assign wavelengths to particles, for in some circumstances they behave just like waves. But quantum theory also tells us that wavelength depends on energy: the higher the energy, the shorter the wavelength. So, the higher the energy of a particle beam, the more detail it will reveal.

The machines planned for the second half of the 1980s will take the smallest distance we can probe down to 10^{-17} cm, or one ten-millionth of the diameter of the atom. To study matter at such small scales requires colossal apparatus, with dimensions of several metres, because it requires very high energies – energies at which the particles in a beam are moving close to the speed of light. The beam may probe a "target" of protons in the guise of a vessel full of liquid hydrogen; or it may intercept another stream of high-energy particles moving in the opposite direction. Either way, the colliding particles will scatter, still moving at very high velocity. Moreover, new particles, often of entirely new varieties, may appear, their mass created from the energy of the collision, in the reverse of the process that fuels a nuclear bomb, where mass transforms into energy. The new particles too may be moving close to the speed of light, travelling several metres in a few hundredths of a microsecond.

The apparatus to detect the products of a high-energy collision must be large enough to provide sufficient information to allow the physicist to re-create the events that occurred, much as we might re-create the scene of a traffic accident. But in high-energy physics the participants leave the scene long before the investigator arrives; the evidence all comes from the equivalent of skid-marks, the trails the minute particles leave as they fly away from the collision.

The basic principles of detecting sub-atomic particles have changed little since the early part of the century, when Ernest Rutherford and his contemporaries began to unlock the mysteries of the atom. Rutherford might not recognise the elaborate apparatus of today, but he would appreciate the methods by which it works.

When an electrically charged particle, such as a proton or an electron, moves through matter, it ionises atoms along its path, losing energy as it does so. The energy goes to release electrons from orbit around their parent atoms, leaving behind positive ions. The amount of ionisation depends on a number of factors – the particle's energy, its mass, its charge – and provides an ideal indicator of the nature of the ionising particle. More massive particles produce more ionisation; those moving at higher velocities produce less, except very close to the speed of light, when their ionising effect begins to increase again.

One of the simplest ways to detect ionisation is to collect the ions (or electrons) at a negative (or positive) electrode. This is particularly simple in a gas, in which the ions and electrons can move quite rapidly. The charge collected provides an electrical pulse that can be made to swing a needle or "click" a counter. This is the principle behind the renowned Geiger counter – a gas-filled tube with a central wire as one electrode, and an outer cylinder as the other. A single detector such as a Geiger counter reveals but one point on the path of a single charged particle. The apparatus built for a large modern accelerator is designed to show the tracks of many particles, often more than 100 at a time, and to provide a detailed picture of the nature of the particles. To this end the equipment consists of a variety of detectors, each designed to record information about different characteristics of the particles, which can all be pieced together "off-line", with the aid of a computer (or two or three!) once the experiment is over. The box shows how

---

Box 31.1   How to build a collider-detector

The basic apparatus at a collider consists of layers of detectors wrapped round the beam pipe at the point where the two beams intersect. The first layer may be a "vertex detector" designed to pinpoint accurately the "vertices" where very short-lived particles decay into others, often still within the beam pipe.

The second layer is generally the main tracking device, recording the paths of charged particles over a distance of a metre or so. It sits within the coils of a large electromagnet, the field of which provides a force on charged particles, bending their paths according to their momentum; particles with least momentum are bent the most. The curvature also reveals the charge on a particle, as the force acts in opposite directions on positive and negative charges.

The third layer of the assembly is often an "electromagnetic calorimeter", designed to trap and measure all the energy carried by electrons and photons. This layer can thus separate electrons from other negative particles, and reveal photons produced in the decays of neutral particles which leave no tracks in the second layer. The fourth layer captures the "hadrons" – those particles that interact with matter through the strong nuclear force, such as protons, neutrons, pions and kaons. This is the "hadron calorimeter", and it generally incorporates the iron that contains the magnetic field.

The only particles likely to penetrate beyond this fourth layer are muons and neutrinos. A final layer of detectors can register the passage of charged particles and so indicate the muons. The uncharged, weakly interacting neutrinos escape completely. Their presence can be inferred only through a detailed accounting of the energy balance in the collisions; missing energy indicates an undetected particle, probably a neutrino.

---

nis works in practice, with an imaginary experiment for a olliding-beam machine. The next generation of accelerators will ll be colliders, with one beam of particles meeting another headn at one or more points around a circular path. This technique eaches far higher energies than when a single beam strikes a tationary target, as Chapter 28 describes.

Each complete assembly for the new generation of accelerators ollows the basic pattern laid down in Box 31.1 (see opposite), with ayers of different detectors forming a concentric "nest" around he collision point. However, this pattern must be tailored to fit in vith the precise parameters of a particular accelerator, and ifferent teams of physicists may choose to accentuate different spects of the collisions they are studying. The final assemblies ary considerably in detail, tackling the same basic problems in

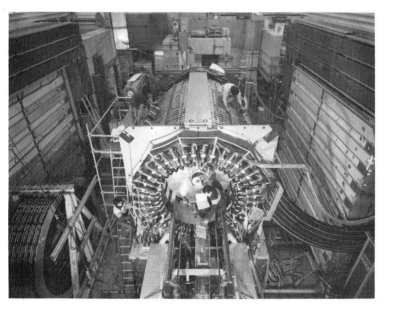

Plate 31.1 *A view of the JADE detector while being put together at the PETRA electron–positron collider at DESY, shows some of its various layers. The large slabs to left and right are the outer muon chambers. The cylindrical part contains the lead–glass to detect electrons and photons, with a central core of tracking chambers surrounding the beam pipe, the end of which is just visible. (Credit: DESY/Schmidt Luchs)*

different ways, while making the most of the latest breakthrough in technology.

The main tracking device is one of the most important parts c the apparatus, providing details about individual particles, whic can later be related to information from other detectors – so as t distinguish between protons, positrons and positive pions, fc instance, all of which have positive charge but widely differin masses. The basic idea is to track charged particles through a larg volume; the general technique relies on the same mechanisms tha underlie the workings of the Geiger counter.

By the 1970s gas-filled chambers had been developed to th extent that the charges could be picked up on an array of parall wires, rather than a single electrode. This has the advantage c providing accurate information about a particle's track, at least i one dimension, corresponding to the position of the wires. Th wire nearest the track produces the biggest signal (Figure 31.1 Chambers with two sets of wires running at different angles giv information in two dimensions, while a third dimension com from having a series of chambers spaced at appropriate interval To obtain good precision the wires in these "multiwire" chambe can be as closely spaced as 1 mm. Imagine the number of wir required to cover a volume of 1 cubic metre, even to provide only few points along a particle's track. The total could easily be man thousand wires, each with its own electrical circuits to amplify an transmit the signals picked up from the ionised trails.

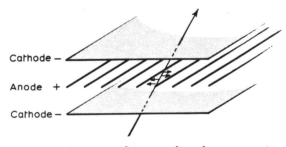

Figure 31.1   *A basic multiwire chamber comprises a plane of closely spaced parallel wires, typically 1–2 mm apart, between two outer planes, generally also formed from wires. The central plane is kept positive relative to the outer planes, and so collects electrons released by the passage of an ionising particle. The wire nearest the particle's track gives the biggest signal.*

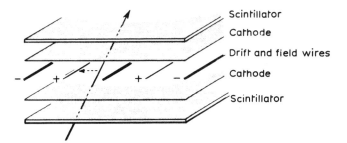

Figure 31.2 *In a drift chamber electrons from ionisation are allowed to move through a uniform electric field to collecting anodes. Their drift time is measured relative to a signal from outer scintillation counters; the time gives an accurate measure of the track's position relative to the nearest wire.*

A variation of the multiwire chamber, known as the "drift chamber", reduces the number of wires to some extent. In this case the electrons released in the ionisation are allowed to drift at constant velocity towards collecting electrodes. The important parameter is the time the electrons take relative to a signal from a separate detector which flags the passage of an ionising particle through the chamber. By recording the time taken with an accurate electronic "stopwatch", the physicist can later calculate the position of the particle's track to great precision – within a few tenths of a millimetre (Figure 31.2). Planes of wires to collect the electrons are generally located every few centimetres, so a few thousand wires can cover a fairly large volume – a cubic metre or so.

Many experiments, both planned and existing, incorporate large drift chambers to track charged particles. But in 1974 David Nygren at the Lawrence Berkeley Laboratory came up with a better idea. He envisaged a single cylindrical chamber, with collecting electrodes only at each end, instead of throughout the whole volume. This would dispense with much of the material in the chamber (which can complicate calculations of the particles' energy) and could reduce the amount of electronics. Conventional drift chambers often bristle with amplifiers and so on, rather like electronic porcupines.

Nygren's brainchild has a suitably space-age name – the time projection chamber, or TPC. The basic idea is to drift the ionisation from a whole length of track through distances of

1 metre or so, to a grid of electrodes. The grid provides a two-dimensional projection of the track; timing the drift provides the third dimension. With a high negative voltage set up on a thin plane across the centre of the chamber, the electrons released by tracks on either side of the plane drift to the nearest end, which is held at zero volts, that is, positive with respect to the chamber's centre. The resulting electric field is parallel to the chamber's axis, and this is the way the electrons drift.

The main difficulty with drifting ionisation over such large distances is that the electrons tend to diffuse away from each other, so spreading out the "track" as it moves towards the collecting electrode. Nygren's answer was to provide a magnetic field parallel to the electric field in the chamber. The magnetic field produces no force in the general direction of motion of the electrons towards the electrode; but it does produce a force that counters the tendency of the electrons to diffuse outwards, and so helps to keep the trail of ionisation intact.

Nygren's TPC at last came to life in 1983 at the Stanford Linear Accelerator Center (SLAC) in California. It forms the central part of an experiment on PEP, a machine that collides together beams of electrons and positrons. The axis of the chamber lies along the vacuum pipe that carries the counter-rotating beams. Particles produced in the collisions spill out into the TPC, leaving trails of ionisation to be swept towards the end plates, where they are duly recorded. The chamber is 2 m long and 2 m in diameter, and is part of an assembly that took five years to put together, involved some 200 people, and cost in the region of $25 million.

The TPC is a basic concept that figures in a number of the proposals for detectors at the new colliders. CERN is building a 27 km circumference electron–positron collider called LEP, which should be ready in 1988. This will reach energies three times as high as those accessible to PEP. Two of the four detector-assemblies chosen to study collisions at LEP will centre on TPCs.

The TPC, and the large-volume drift chambers, provide full three-dimensional images of tracks, which can be reproduced by computer graphics, often in startling colour to show particles of different momentum. But the precision of these devices is limited; SLAC's TPC can measure points to a little better than 200 micrometres. Higher precision is needed than this to sort out exactly what happens in the region close to the initial collision point.

Short-lived particles formed in the collision may soon decay to other particles; the detectors must be able to show the physicists which particles emanate directly from the collision and which come from decays a short distance away. This implies the ability to see whether tracks converge back to a single point, or whether they meet at two or three distinct points. For this purpose, many of the new experiments employ a central "precision vertex detector", based on that most ubiquitous material of the electronics age, silicon.

Silicon has been used to detect ionising particles since the early 1950s, but it has figured mainly in nuclear physics experiments, as opposed to high-energy experiments where large-area detectors are generally required. Now, however, high-energy physicists are rediscovering silicon to provide the accuracy needed in small detectors close to the particle interactions.

A semiconductor, such as silicon, is half-way between being a metal, with good conductivity, and an insulator, with poor conductivity. But it takes a relatively small amount of energy to "release" electrons in a semiconductor from their parent atoms, so that they can move freely round the material – in other words, to convert the material to a conductor. Indeed, the same amount of energy can release ten times as many electrons in silicon as in a gas; in a slice of silicon only 0.3 mm thick, the passage of a charged particle can release more than 20 000 electrons, which are then free to conduct.

To be able to use silicon as a particle detector, the problem is to collect the extra electrons while avoiding setting up a normal electric current, which could swamp the signal due to the particle. The trick is to make a silicon diode – a device that conducts when the electric field across it is in one direction, but not in the other. An electric field can then be used to direct the electrons released by the particle (and the positive "holes" vacated by the electrons) to collecting electrodes, just as in a gas-filled chamber (Figure 31.3).

A team from the Max Planck Institute, the Technical University in Munich, and CERN, has fabricated particle detectors on wafers of silicon 0.3 mm thick, with an area of 24 mm by 36 mm. These researchers used standard techniques to make 1200 diodes on the wafer, in the form of strips each 36 mm long and 0.02 mm apart. Signals are collected only from every third strip, but as the strips are coupled in an electrical sense, information from the intervening strips is not lost.

Munich's silicon-strip detectors have already been put through their paces in an experiment at CERN to look for the decays of

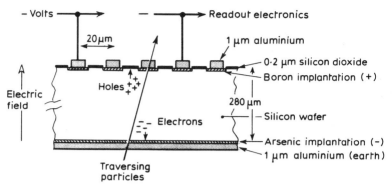

Figure 31.3 *A silicon-strip detector is formed from collecting electrodes – the strips – laid down on a silicon wafer. "Holes", with positive charge, released by a traversing charged particle, are collected by the nearest strip.*

the so-called "charmed mesons". These particles live for only $10^{-12} - 10^{-13}$ s before decaying, and travel only a few millimetres or so in their brief lives. But with the precision available from the silicon detectors, the experiment could resolve particle tracks only 0.1 mm apart.

Two of the new detector assemblies at LEP plan to use planes of silicon-strip detectors wrapped round the beam pipe to provide precise information as close to the collision point as possible. One proposed experiment is contemplating a total of more than 100 000 strips encircling the beam pipe on three cylindrical supports, each 42 cm long. Nearly half-way round the world at SLAC, silicon detectors also have a role to play at the new high energies.

Thanks to a cunning modification of the existing 3-km long linear accelerator, SLAC may produce electron–positron collisions of the same energies as at LEP but even sooner. A pulse of positrons will be sent down the linac almost imediately after each pulse of electrons; at the end the two sets of particles will be directed oppositely round the same circle to meet head-on, half-way round (see Chapter 29). One experiment proposed for this new facility will incorporate silicon detectors at its heart. But there is another proposal to go one better than silicon-strip detectors, and that is to use another product of the microelectronics age – the CCD or charge-coupled device.

CCDs already play an important role in astronomy, where they ːe used in very sensitive cameras to pick up minute levels of light. he image is recorded in a two-dimensional array of "pixels", ɪch about 0.02 mm square, formed on a silicon wafer a square ɪntimetre or so in area. Each pixel is in fact an area of the chip ʿhere charge – electrons – can collect. The charge can be roduced by incident light, as in astronomical applications, or by ɪe passage of a charged particle, just as in the silicon-strip ɛtectors. By adjusting the relative voltages on the pixels, the buckets" of stored charge can be made to move along the rows of ɪe two-dimensional matrix, and then fed sequentially into ɛctronics that can analyse and record the information.

The CCD offers a basic advantage over silicon strips: two-ɪmensional information from a single device, which in turn lows for a better separation of closely spaced tracks. Imagine, for ɪstance, tracks spreading out in one plane like a fan. Viewed from ɪe direction the tracks will be clearly separated, but from a ɪrection at 90° to the first, they will be more or less superimposed. single silicon-strip detector – like a simple multiwire chamber – ʋes a view in only one direction; two views at right angles must ː combined to reveal the tracks in space. The CCD provides two-mensional information directly, and its inherent precision is ɛtter reflected in its ability to resolve closely spaced tracks.

There are, however, two difficulties with CCDs. First, the chips ɪust be operated at low temperatures, around 120 K; at room ɪmperature too many electron–hole pairs are created thermally, ɪd these swamp any signal due to the passage of a charged ɪrticle. Secondly, it takes a long time – several tens of milli-conds – to read the information out of a CCD, whereas at an ːcelerator the successive bursts of particles generally arrive every ʋ hundredths of a millisecond. However, undaunted, a group ɔm the Rutherford Appleton Laboratory succeeded in testing CD detectors in a high-energy physics experiment at CERN, ʋards the end of 1983.

The CCD particle detector may come into its own at the SLAC ɪnear Collider (SLC). There the two beams of particles are ɟueezed down to a few micrometres across before colliding, so ɪall CCD detectors can be arranged around the collision point. ɪoreover, the frequency of the collisions can be more easily ɪjusted in this machine, to match the relatively slow read-out of e CCDs.

A third type of silicon-based detector is so new that the idea is

Plate 31.2 *A test rig shows a CCD detector at the heart of the cryostat needed to keep it at its operating temperature of about 120 K. The inset shows the CCD pixels that have "fired" with the passage of 17 charged particles. (Credit: CERN/RAL)*

not yet tested. This is the silicon drift chamber, which combines the principles of silicon detectors with those of the gas-filled devices. Invented by Emilio Gatti, from the Polytechnic of Milan, and Pavel Rehak from the Brookhaven National Laboratory in New York, the device is basically a silicon wafer drained of all its normal free charge-carriers – a state referred to as "fully depleted". Electrodes in the form of metallised strips on both surfaces provide a field that directs any electrons released by a traversing particle towards a collecting electrode at one end. The advantage of such a detector is that it converts information on *positions* into *times*, and signals from a relatively large area can be picked up by a single electrode. This would overcome some of the tremendous difficulties involved in simply taking the signals off the basic chips, and transmitting them for combination with signals from other parts of a large assembly of apparatus.

A knowledge of the precise tracks of particles and the locations at which they were formed or where they decayed, is only part of the story. It is also important to be able to identify the different varieties of particle produced. Nygren's TPC attempts to combine both tasks: tracking and identification. However, it is often more efficient and more convenient to use different detectors to provide the complementary pieces of information. The Čerenkov effect, a phenomenon known for 50 years, is being updated to identify particles in many of the new detector assemblies, both at LEP and the SLC.

In the 1930s, Pavel Čerenkov discovered that if charged particles move through a substance at a velocity faster than light *in that substance*, then they produce an electromagnetic "shock wave" – in other words, they emit a cone of light about the path of the particle. This effect is possible only because substances like glass and water slow down light, so that particles can indeed move faster through them than does light; but note that the particles do not exceed the limiting velocity of light *in a vacuum*.

A simple Čerenkov detector registers only whether or not a particle produces light; that is, whether its velocity is above or below a certain value. It thus distinguishes particles such as protons and pions, which can have the same momentum (mass × velocity), but widely different velocities due to their large difference in mass. The new variation on this theme, sparked off in 1977 by Tom Ypsilantis and J. Séguinot, is more sophisticated. Their "ring imaging Čerenkov counter", or RICH, aims to detect the cones of light in space, for a whole range of particle directions.

Thus it can measure a particle's velocity according to the exact angle of the cone, determined from the radius of a *ring* of light intercepted by a two-dimensional photon detector. The location of the centre of the ring depends on the direction of the particle. The art in building a RICH detector lies in catching the small numbers of photons in the Čerenkov radiation. One of the first methods investigated was to use a hydrocarbon vapour to "convert" the Čerenkov photons to electron–ion pairs. The electrons could then be detected by a multiwire proportional chamber, with some form of two-dimensional read-out.

The largest working RICH detector has been built mainly by a team from the Rutherford Appleton Laboratory, and the Universities of Lancaster, Manchester and Sheffield. It is installed at CERN in an experiment that studies the interactions of a secondary photon beam generated initially by particles from the 450-GeV Super Proton Synchrotron. A curved array of 80 hexagonal mirrors, 7×4 m^2 in area, focuses the Čerenkov light from charged particles travelling through a volume of nitrogen back on to a bank of wire chambers. These wire chambers are in fact small TPCs, filled with a mixture of methane, isobutane and TMAE (Tetrakis (dimethylamine) ethylene). The TMAE converts the Čerenkov photons to electrons.

The mirrors had to be specially made in the UK by A. E. Optics in Luton, and were coated at the Culham Laboratory to make them good reflectors of the far ultraviolet wavelengths of the Čerenkov radiation. The detector observed its first rings last October. It can distinguish pions from protons at energies as high as 160 GeV, thanks to the ability of the TPCs to reproduce the positions of the Čerenkov photons to an accuracy of 2.6 mm.

Barry Robinson and colleagues at the University of Pennsylvania have taken a different approach in a RICH developed for use at the Fermi National Accelerator Laboratory in Illinois. Spherical mirrors focus the Čerenkov light to form rings, which are intercepted by an image intensifier. This produces a spot of light for each photon originally detected, which is in turn picked up by photo-electric detectors.

In yet another variation on the RICH theme, built by researchers at the Serpukhov Institute for High Energy Physics in the USSR, the rings of light are brought to focus outside a cylindrical gas-filled Čerenkov detector. The rings are intercepted by tubes with 20-cm long photo-sensitive cathodes, which convert the photons to electrons. The time of the phototube's output signal depends on the exact position at which the photon strikes the cathode.

The detectors at the new colliding beam machines will make the most of modern technology. But what of the old faithful bubble chamber? Bubble chambers are not suited to colliders, but they still have an important role to play at "fixed-target" machines, where particle beams are extracted and directed at a target. And technological innovation is invading the province of bubble chambers, too, as they are directed into searches for the short-lived charmed particles.

The problem that confronts physicists studying charmed particles is that they are unstable and are very short-lived. A neutron outside an atomic nucleus lives an average of 15 minutes before it disintegrates into a proton, an electron and a neutral particle called a neutrino; a charged π-meson lasts only 10^{-8} s or so before decaying into a muon (a sort of heavy electron) and a neutrino. But charmed particles last for less than 10^{-12}s, or a million-millionth of a second.

The short lifetimes of charmed particles means that they are difficult to pinpoint in the debris of high-energy collisions. Generally, physicists must look for combinations of other, longer-lived particles that can have come from the decay of a charmed particle. This procedure is made much simpler if an experiment can resolve the point at which a charmed particle decays separately from the position at which it was created. The particle will travel at most only a few millimetres during its brief lifetime which makes things difficult. But the task is made even more difficult by the fact that during this short space of time the charmed particle does not move far from the so-called "forward" direction – the direction defined by the path of the high-energy particle that initiates the collision, and in which most of the debris tends to travel. Indeed, it is by looking for the point from which the tracks of the decay products appear to emanate that researchers can pick out a decaying particle, as Figure 31.4 shows. For a lifetime of 10^{-12} s calculations show that a particle can stray no more than about $300\,\mu$m from the forward direction, so any means of recording the point of origin of the decay products must be able to resolve dimensions of $100\,\mu$m or so.

And this is where the trouble begins for bubble chambers. What *is* the smallest size of bubble that you can photograph? To record the production and decay of a charmed particle requires a

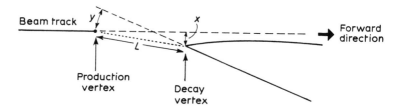

Figure 31.4 *Charmed particles live for only 10^{-10} s, so the critical dimension y is typically less than 300 μm, although L may be up to a few millimetres. To reveal the production and decay vertices a track detector must be able to resolve distances smaller than y — or 30 μm or less in the case of charmed decays.*

resolution of 100 μm or better, if the point of decay, or "vertex", is to be picked out from the complex pattern of tracks. In this case the bubbles can be allowed to grow only to 30 μm or less, before being photographed. The difficulty that arises here is that the camera's ability to resolve detail is related to its depth of field, D – the range that is in focus. Indeed, the resolution equals $0.6\sqrt{\lambda D}$ where λ is the wavelength of light. Thus if an optical system is to resolve tiny bubbles, it can observe only a small volume – a problem familiar to photographers and anyone who has used a microscope. For these reasons, Colin Fisher, at the UK's Rutherford Appleton Laboratory (RAL), suggested in 1978 using small rapid-cycling bubble chambers for the direct detection of charmed-particle decays.

In the following months, Heinrich Leutz and his team at CERN built LEBC, the Lexan bubble chamber, which has a diameter of 20 cm, is 4 cm deep, and holds 1.1 litres of liquid hydrogen. Lexan is a transparent plastic which behaves well mechanically at the low temperatures required to keep hydrogen a liquid. The bubbles produced by high-energy particles were photographed when about 40 μm in diameter, only 300 microseconds or so after the chamber's expansion. The corresponding depth of field was around 5.4 mm, which is sufficient to contain the incident beam of particles used in the tests, but close to the smallest practical limit.

A team of experimenters from many European institutions used LEBC to study the interactions of high-energy negative π-mesons from CERN's largest accelerator. The high resolution achieved with LEBC allowed the team to sort out those pictures, or

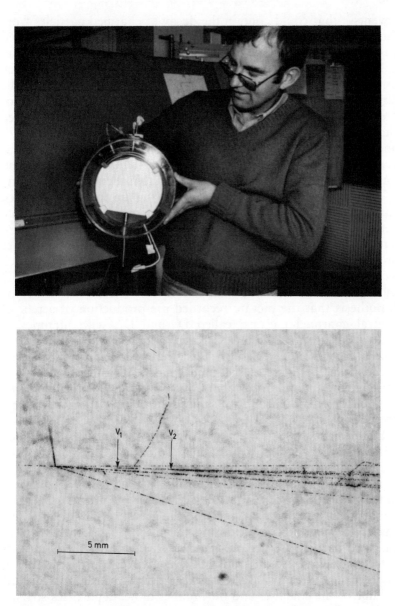

Plate 31.3 *LEBC, the Lexan bubble chamber, was specifically built to look for charmed particles; in the "event" shown here two neutral charmed particles decayed into charged secondaries at V_1 and V_2. (Credit: C. Sutton; CERN/ RAL)*

"events", that could contain the decay of charmed particles from those containing the decays of other particles. Knowing that particles with a lifetime of 10^{-12} s will spread no further than ± 300 μm from the forward direction, the physicists could search for decays that occurred within a restricted volume around the direction of the beam particles. In particular they searched for events in which they could see two decays, as particles with naked charm are expected to be created in pairs, such that the charm of one particle is cancelled by the anti-charm of the other, thus maintaining a net charm of zero. Out of 48 000 events, the team found 20 cases with a pair of decays within their chosen region.

To prove that they were indeed observing the decays of charmed particles, the team repeated their experiment using additional detectors – the 40-metre long European Hybrid Spectrometer – outside LEBC to measure the momentum of charged particles produced in the decays and to detect neutral π-mesons which would not leave tracks in the bubble chamber. The Plate 31.3 shows an event in which the data are consistent with the hypothesis that the picture recorded the production of a pair of neutral charmed mesons, called D^0 and $\bar{D}^0$, with lifetimes of 2.1×10^{-13} s and 5.9×10^{-13} s respectively (*Physics Letters*, vol. 102B, p. 285). These results were particularly intriguing because the values are several times longer than those measured in earlier experiments using emulsion techniques, and encouraged physicists to try harder to achieve high resolution in bubble chambers.

Experiments such as the one with LEBC show that conventional optical systems can have sufficient resolution to separate out the decays of particles with lifetimes in the range 10^{-12} to 10^{-13} s, so is there any need to improve the optics still further? The answer to this is yes, for two reasons. First, it would be helpful to increase the depth of field, so as to increase the useful volume of a bubble chamber, without compromising the resolution. Other types of charmed particle are not produced as frequently as the D-mesons detected in LEBC, so the study of these involves a larger number of interactions. The most efficient way of doing this is to increase the volume of the chamber and the number of beam particles striking it. Secondly, with improved resolution, bubble chambers could reveal particles with still shorter lifetimes than those of the charmed particles. Particles containing a fifth type of quark, known as bottom, might live for only as little as 10^{-14} s or so, and to observe directly the decay of such short-lived objects will require a resolution in the region of 5–10 μm. But with conventional optics the corresponding depth of field would be less than 0.5 mm, too shallow to be useful.

One way round the problem of the trade-off between resolution and depth of field is to use holography to record images of the trails of bubbles. The basic principle of holography is to "freeze" on emulsion the interference pattern between light scattered from an object and light in a reference beam. The pattern produced bears little or no resemblance to the object, but it stores all the information carried by the light from the object. The image of the object is recreated when the hologram – the developed plate or film – is illuminated with light identical to that used to create it. A hologram can be recorded only with coherent light – that is, when the waves are perfectly in step – and for this reason the pure light from lasers is used.

The idea of using holography in bubble chambers first came in 1966, from Walter Welford at Imperial College, London; it was the late 1970s before Colin Fisher at RAL, Paul Lecoq at CERN and Fred Eisler at the City University of New York revived the idea to solve the problem of detecting charm and beauty. The value of holography is that it allows high resolution at the same time as a large depth of field. With conventional optics the depth of field is related to the resolution by the equation $D = 3\lambda(r/\lambda)^2$; with holography the equivalent relation becomes $D = 0.08L(r/\lambda)^2$, where L is a property of the laser called the coherence length and can be as large as a few metres. Thus for the same resolution, the depth of field of a holographic system is much greater; for a resolution of a few micrometers it can in principle be several metres.

The first tests with holography in a bubble chamber were performed at CERN during the summer of 1980 with a small chamber containing Freon. Freon is sometimes used in preference to hydrogen because it is a liquid at much higher temperatures and is therefore far easier to handle; but it presents a much more complex "target" than hydrogen, with nuclei of many protons bound together with neutrons, and this makes the pictures more difficult to analyse.

The chamber, called BIBC for Berne Infinitesimal Bubble Chamber, was only 6.5 cm in diameter and had been used with conventional optics to look for charmed-particle decays. Its two windows made the chamber particularly easy to adapt for use with "in-line" holography. In this technique the same laser beam illuminates the bubbles, and acts as reference beam. The team could shine a laser beam across BIBC, at right angles to the tracks of particles, to form a hologram on a glass plate, or a polyester film, coated with photographic emulsion. The tests showed that bubbles as small as 8 μm could be recorded successfully, the pulse of the laser beam coming only 7 μs or so after the particles had passed through the chamber.

While the tests with BIBC were under way, groups at CERN and RAL began to design small chambers specifically for use with holography. Two chambers have now been built and tested, one containing hydrogen, the other Freon. HOBC, for holographic bubble chamber, was built at CERN by Alain Hervé. Like BIBC it contains Freon, and in a beam of particles it has produced holograms with a resolution of 10 μm as shown in Plate 31.4. The hydrogen bubble chamber HOLEBC (holographic Lexan bubble chamber) has been designed, again by Leutz at CERN, to use with the huge European Hybrid Spectrometer to identify and measure the momentum of particles as they emerge from the chamber. In contrast to the original LEBC, the holographic version has two windows, so that holograms can be made. This changes the way the camera's flash in a conventional system illuminates the bubbles, and HOLEBC has photographed classically bubbles only 20 μm across. Tests have shown that with holography the chamber can resolve bubbles as small as 10 μm, and tests with fine wires show that even better resolution is possible in principle.

The large depth of field that holography can in principle achieve has encouraged physicists to consider modifying the large bubble chambers, such as BEBC, that are used with neutrino beams. Although interactions between neutrinos and matter are much rarer than those initiated by protons or π-mesons, experiments with neutrinos provide a "cleaner" means of studying charm. Whereas charmed particles are produced in only 0.1 per cent of proton or meson interactions, they appear in some 10 per cent of neutrino events. Experiments with neutrinos also offer the possibility of searching for decays of other short-lived particles, in particular the so-called τ-lepton, an unstable heavy relative of the electron, discovered in the mid-1970s.

Holography would seem to provide the only means of achieving the fine resolution needed to observe charmed-particle decays, while maintaining a reasonably high rate of neutrinos incident on a large bubble chamber. Groups at RAL, CERN and the Institut St Louis in Mulhouse have tested laser systems set up on a bench

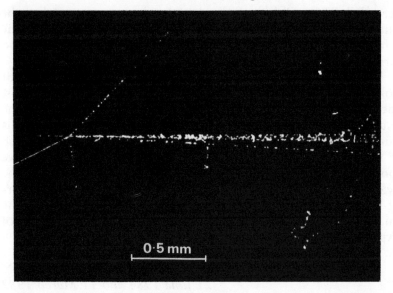

Plate 31.4 *An interaction reconstructed from a hologram taken in the chamber HOBC. The bubbles are 12 μm diameter. (Credit: CERN)*

to reproduce the vast size of BEBC. These tests have already successfully created holograms of wires a few tens of micrometres in diameter. Groups in the US are also considering how to modify the 15-foot (4.5-m) chamber at Fermilab, the national accelerator laboratory in Illinois, to take holograms of bubble tracks.

Holography in bubble chambers may well turn out to be one of the first uses on a large scale of a technique that is perhaps more familiar for its use in novelties, or in advertising displays. A single experiment with either a small-scale or a large-scale chamber will produce some tens of thousands of holograms.

The analysis of so many holograms is just one problem that the physicists have to solve, and several groups are already working out how to adapt the present systems used for automatically scanning photographs taken in bubble chambers. Although a hologram will have a large depth of field, the optical system used to look at it when it is "replayed" will still be limited and will be able to scan only a thin slice at a time. Some way of knowing where in the field of the hologram the interesting event took place would be useful, as well as some way of ensuring that only holograms of the few per cent of potentially-interesting interactions are recorded. This will have to be done by detectors external to the bubble chamber, which can respond rapidly enough to provide the information in the few microseconds before the bubbles grow too large.

The holographic bubble chambers illustrate well the way that "old" techniques can proceed through a number of reincarnations, each brought to life by technological innovations. But as the examples of the TPC and the RICH detectors also show, it takes a number of years for a particle detector to reach the stage at which it becomes a common feature of experiments. Here we have looked at some of the techniques available now, for use on the next generation of accelerators in the late 1980s. The detectors of the 1990s must already be on the drawing boards in the imaginations of Rutherford's heirs – the latest generation of experimenters.

CHRISTINE SUTTON
21 June 1984 and 11 March 1982

EPILOGUE

Is there physics after the Z^0?

JOHN ELLIS

The discoveries of the W and Z particles showed that particle physicists seem to be on the right track in understanding the particles and forces that make up our Universe. But there is still much work to be done.

Our understanding of particle physics has advanced rapidly over the past few years, as the previous chapters have shown. Such progress has led to a broad consensus among particle physicists on the nature of the correct "standard model" of the fundamental interactions between particles. Does this mean that there is no need to put more people and money into particle physics, and that all physicists on active duty should be redeployed as computer-game designers (or whatever), after handing out a few Nobel prizes as medals for the heroes of the campaign? Far from it, as we will see shortly.

Many questions of principle arise within the context of the "standard model", whose answers are crucial not only to our understanding of the fundamental interactions, but also to the developing interface between particle physics and cosmology. At a more detailed level the "standard model" has many untested features and is inadequate in many respects. Different theorists propose different solutions to these problems, which can only be resolved by experiment. I would like to describe some of the crucial experimental tests to be carried out in the coming years. The next decade in particle physics may well be as exciting as the past decade has been.

Particle physicists distinguish four fundamental interactions, namely gravitation, electromagnetism, the weak forces responsible for radioactivity, and the strong forces that bind nuclei. The basic building blocks of matter on which these forces act are "leptons", such as the electron and the neutrinos, which do not feel the strong nuclear forces; and the "quarks", which are the basic constituents of strongly interacting particles (hadrons) such as protons and neutrons. One of the great advances in particle physics during the past few years has been the discovery of these underlying quark constituents of the strongly interacting particles. Equally important has been our development of an understanding of the basic dynamics of nuclear forces, which are now known to be based on a theory called quantum chromodynamics (QCD). This theory demands eight massless "gluons", to carry the strong forces between the quarks, which are believed to become so strong at "large" distances – in the region of 10^{-15} m – that the quarks and gluons are confined to the interiors of hadrons such as the proton, neutron and mesons.

The gluons have one unit of intrinsic angular momentum (they have spin 1), just like the photon of quantum electrodynamics (QED), the quantum theory of electromagnetism. The basic particles of matter between which the gluons and photons carry forces, namely the quarks and leptons, also have intrinsic angular momentum, in their case half a unit (spin ½). It has long been known that QED has a powerful symmetry called "gauge invariance", which allows the mathematical descriptions of the quarks and leptons to be changed independently at each point in space–time. QCD has an analogous "gauge symmetry", whose presence is necessary if the interactions of the gluons and of the photon are to be theoretically consistent. Both QED and QCD are thus "gauge theories".

In parallel to this progress in understanding the strong nuclear forces there has been dramatic progress in our understanding of the weak interactions and of their unification with electromagnetism. The discoveries of the $W^{\pm}$ and Z^0 particles at CERN have confirmed that the weak interactions are carried by massive spin 1 particles, just as predicted by the original unified gauge theory of electroweak interactions proposed by Sheldon Glashow, Abdus Salam and Steven Weinberg. This theory together with QCD constitutes the "standard model" of elementary particles and their interactions. Meanwhile, the fourth fundamental interaction, gravity, is still believed to be well described by Albert Einstein's theory of general relativity, at the energies and distances presently accessible in experiments.

The general acceptance of this "standard model" certainly does not mean that physics has no more important questions to be answered after discovery of the $W^\pm$ and Z^0, any more than the experimental discovery of the electromagnetic waves predicted by James Clerk Maxwell meant that physics was finished a century ago. (Some physicists thought so at the time, but that is not an error we should repeat.) In that case, progress was subsequently made in two directions. On the one hand, attempts to unify or reconcile the classical theories comprising the "standard model" of the time – namely Newtonian mechanics and Maxwell's electromagnetism – led to the theory of relativity. On the other hand, attacks on niggling problems in statistical physics outside the scope of that "standard model" led to quantum mechanics. Perhaps progress can now be made in two analogous directions: by attempts to unify the fundamental gauge interactions in a more elegant and compact form, such as grand unified theories, and by attacks on niggling problems which are beyond the scope of present gauge theories. Why, for instance, are the masses of particles what they are? And why are there so many types of quarks and leptons?

The development of quantum mechanics out of statistical physics may have caused a more revolutionary change in the way physicists tried to describe reality than did the emergence of relativity. It led to the abandonment of causality and determinism, which were replaced by "God playing dice". Thus it may be that attacks on problems beyond the reach of gauge theories may bring more revolutionary change in the near future. At least some of the theories attacking these problems are also more subject to experimental test in forthcoming accelerator experiments, than are the grand unified theories, so let us concentrate on them in this chapter.

Gauge symmetry is a potent unifying feature of the different fundamental interactions: strong, electromagnetic and weak with associated spin 1 "bosons" – the gluons, photon, $W^\pm$ and Z^0 respectively. However, there is one striking difference between the weak bosons and the others, namely that they have masses of 80 to 90 times the proton mass (80–90 GeV), whereas the gluons and photon are apparently massless. To give masses to the $W^\pm$ and Z^0, we must break the gauge symmetry in some way. If the theory is to provide a basis for reliable calculations, the gauge symmetry must be "spontaneously broken" as proposed by Salam and Weinberg.

This requires the existence of a field without intrinsic angular momentum (spin 0), which takes a non-zero value throughout space–time. This spin 0 field is called the Higgs field, and associated with it there should be physical particles called Higgs bosons. The existence of one or more Higgs bosons with an unknown mass less than about 1 TeV (1000 GeV) appears inevitable in any reliable electroweak theory. It is therefore vitally important to plan experimental searches for such a Higgs boson. The discovery of the $W^\pm$ and Z^0 seem to confirm the relevance of the gauge principle: now we must find out how it is broken.

The best prospects for experimental searches for Higgs bosons may be provided by high-energy electron–positron (e^+e^-) collisions. The Glashow–Salam–Weinberg theory requires a sixth type of quark (called "top"), which is confined inside hadrons as are its lighter quark brethren. Bound states of a top quark with a top anti-quark – "toponium" – will be produced in e^+e^- collisions at sufficiently high energy. Theoretical estimates suggest that if there is an electrically neutral Higgs boson whose mass is less than that of toponium, then we should be able to observe toponium decaying into a photon and a Higgs. Unfortunately, we have no good estimate of the mass of either toponium or the Higgs boson. It is possible that the Higgs may be too heavy to be produced by decaying toponium, and other experiments to search for the Higgs should be kept in mind.

If it has a mass less than about 50 GeV, the Higgs boson may appear in the decays of Z^0 particles. If it is heavier, the Higgs may only be produced at very high energies, in association with a Z^0. Such experiments to search for the Higgs particle are at the top of the agenda at the Large Electron–Positron Collider (LEP), being constructed at CERN. If the Higgs boson is very heavy indeed, it may be that one needs a high-energy proton–proton or proton–anti-proton colliding beam accelerator to produce it. The present proton–anti-proton collider at CERN is unlikely to be able to produce Higgs bosons in detectable numbers, but it may be possible at the next generation of colliders now being discussed in the United States and Europe. The problem will be to detect the Higgs among all the other particles produced in less interesting collisions. While hunting the Higgs is a top priority in experimental physics, the uncertainty in its mass means that there are no guarantees that any individual experiment can find it.

Although the Higgs boson is an essential feature of the Salam–Weinberg model, being associated with the acquisition of mass by the quarks and leptons as well as by the $W^\pm$ and Z^0 bosons, the unknown mass of the Higgs itself presents severe theoretical problems which raise profound questions about its nature. The mass of any elementary spin 0 particle is extremely unstable. In particular, "quantum" corrections to the masses of spin 0 particles due to interactions with virtual particles, borrowed as it were from the energy of the underlying fields, turn out to be infinite. One can make them finite only by cutting them off in an *ad hoc* manner, which may mimic new physics.

Calculations in gauge theories are generally so well-behaved that there is no need to invent new physics at low energies, and it is often suggested that no new physics may appear much below the scale at which quantum gravitation becomes important, at energies around 10^{19} GeV. If such a large scale is taken as the cut-off to make the quantum corrections finite, then the corrections are also large. Indeed, they are much larger than the mass of the Higgs boson of 100 GeV or so that is required if the approach of Salam and Weinberg is to be consistent. This difficulty is often called the "hierarchy problem". It has persuaded many theorists that an elementary Higgs boson is intellectually disreputable, and they are casting around for alternatives.

One possibility is that the Higgs boson is not an elementary spin 0 particle at all, but is actually composed of spin ½ constituents, much in the same way that the pi-mesons (which have zero-spin) are built from quarks and anti-quarks. Quarks and anti-quarks are bound together by the strong nuclear forces, but to make composite Higgs bosons one must postulate a new set of strong forces, which bind together a new set of spin ½ constituent particles. These new spin ½ particles are often called "technifermions" – where "fermions" refers to the fact that they are spin ½. The new strong interactions are often called technicolour.

Conventional strong interactions have a range of about 10^{-15} m and yield bound states such as the proton and neutron, which have masses around 1 GeV. The new technicolour interactions, on the other hand, should have a range about a thousand times smaller – around 10^{-18} m – and most of the bound states of technifermions should have masses starting around 1000 GeV, or 1 TeV. However, there will also be some lighter particles: the composite spin 0 "technipions" which replace the Higgs boson required by other theories and which have masses much less than 1 TeV, just as the conventional pions have masses much less than 1 GeV. Thus, technicolour theories seem to promise "jam today", in the form of light technipions which could be produced with today's accelerators; and "jam tomorrow", with a complete new spectrum of "technihadrons", which would be at least as rich as the spectrum of hadrons that kept particle physicists busy during the 1960s, and which could be explored with the next generation of accelerators.

Searches for the "light" technipions have already begun at existing accelerators, and it is an embarrassment for some technicolour models that no electrically charged technipions have yet been found in e^+e^- collisions. Some technicolour models also have problems reproducing certain effects of the weak interactions. Because of these two difficulties, and because no complete, elegant and convincing technicolour model has yet been constructed, interest in technicolour theories has declined recently. However, it would be premature to abandon technicolour, which was a very attractive idea in its original form, and experiments should continue to search for technipions in high-energy e^+e^- and hadron–hadron collisions.

Many of the theorists previously interested in technicolour have now turned their attention to another possible resolution of the problems associated with Higgs bosons, namely the ideas of supersymmetry (SUSY). In contrast to technicolour, which made Higgs bosons composite, the idea here is to work with elementary spin 0 bosons and to postulate a new symmetry which protects them from acquiring large masses through the quantum corrections. The key observation is that the quantum corrections involving bosonic (integer spin – 0, 1, . . .) and fermionic (half-integer spin – $\frac{1}{2}$, $\frac{3}{2}$, . . .) particles have opposite signs. Therefore, a suitable symmetry between the fermions and bosons could cancel out the offending quantum corrections. This is just what SUSY provides, since it requires bosons and fermions to exist in pairs with identical interaction strengths and similar masses, if the supersymmetric is not too badly broken.

What does that mean in our particular case? I have already mentioned that for reasons of consistency in the Salam–Weinberg model the Higgs boson cannot weigh much more than the $W^\pm$ particle does, say around 100 GeV. For the quantum corrections to the mass of the Higgs boson to be no larger than 100 GeV the supersymmetry fermion and boson partners cannot have masses differing by much more than 100 or 1000 GeV. This practical argument suggests that the supersymmetric particles cannot be far beyond our present experimental reach, and it is natural to ask whether any of them have been seen already.

Unfortunately, none of the known particles can be the supersymmetric partner of any other known particle. Pairs of super-symmetric partners must have spins differing by half a unit; for instance, spin ½ particles are paired with spin 0 or spin 1. We have seen no spin 0 particles to go with the quarks and leptons, and the known spin 1 particles are gauge bosons, which do not have the same internal properties as the quarks and leptons. Therefore, supersymmetric theorists are forced to invent a doubling of the known spectrum of particles, and lots of new names for the unseen (obscene?) "sparticles", such as the spin 0 super-symmetric quarks and leptons ("squarks" and "sleptons"), spin ½ partners of the $W^{\pm}$ and Z^{0} ("wino" and "zino") and so on.

This doubling of the elementary-particle spectrum is unpalatable to some physicists, but is a natural extension of the same logic that led theorists to propose the existence of charm, and of the W and Z particles in the first place. In all these cases it was realised that quantum corrections would destroy what seemed at first sight

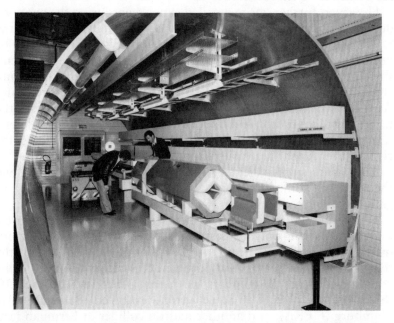

Plate 32.1 *A model of the tunnel for the LEP collider being built at CERN. Electron–positron collisions in LEP at a total energy of 100 GeV may reveal the Higgs particle, required by the standard model of electroweak unification. (Credit: CERN)*

to be an acceptable theory, and the answer was to postulate new particles which tamed the quantum corrections. It is often said that nature does not economise on particles, but on principles. Supersymmetry, or SUSY, is such an elegant principle that surely one should not begrudge her the proliferation of new particles that she entails.

It is an encouragement to experimenters that so many new particles should be lurking "just around the corner", accessible to forthcoming experiments, and possibly waiting to be found at existing accelerators. The lightest "sparticle" may well be the photino (the spin $\frac{1}{2}$ spartner of the photon) and it could well be light enough to be produced in pairs by e^+e^- annihilation at present energies. Another possibility is that the $W^\pm$ and Z^0 may be able to decay into supersymmetric particles. With luck, these could be detected at the existing proton–anti-proton collider at CERN, and experimenters are eagerly scanning their collider data for evidence of supersymmetry.

However, further, more detailed experiments using the many Z^0 decays obtained from higher energy e^+e^- colliders, such as the SLAC Linear Collider (SLC) at Stanford or LEP at CERN, may be needed to find SUSY. Even these devices may not be sufficient, since sparticles could well weigh up to 1 TeV (1000 GeV), and colliders able to reach 1 or 2 TeV total energy may be necessary. At the moment it seems easier to attain such energies using the technology of proton–proton or proton–anti-proton colliders, rather than using e^+e^- colliders, though the latter may provide cleaner events that are easier to interpret.

This discussion indicates that there is a consensus among theories that some new physics associated with the breaking of gauge symmetry should be awaiting discovery at energies up to 1 TeV, but there is no consensus as to what it is. Theoretical imaginations can only be restrained by new accelerator experiments in the energy range up to 1 TeV. For this reason it is healthy that a diversified set of new particle accelerators are now being constructed – e^+e^- colliders in Japan (TRISTAN), at SLAC (SLC) and at CERN (LEP); an electron–proton collider at DESY in Hamburg (HERA); and another hadron collider at Fermilab near Chicago, to join the already successful CERN proton–anti-proton collider.

There is a widespread perception in the United States that Europe is now leading the way in particle physics, particularly in the wake of the $W^\pm$ and Z^0 discoveries. However, no true red-

Plate 32.2 *The electron–proton collider, HERA, being built at DESY, will accelerate electrons to 30 GeV and protons to 820 GeV before bringing the beams to meet head-on. This is the first superconducting bending magnet built for the proton ring. (Credit: DESY)*

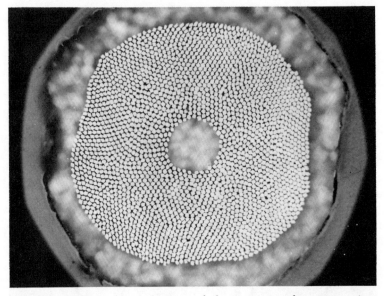

A cross-section through one of the 0.8-mm diameter wires that make up a superconducting cable in the magnet coil. Machines such as HERA will guide theory in the coming years. (Credit: DESY)

blooded American could tolerate such a situation for long, and there began in 1983 a great drive in the American physics community, encouraged by President Reagan's science adviser George Keyworth, to go one better than CERN and be the first into the crucial 1 TeV energy region. The route to this goal that is being most actively discussed is to build an enormous proton–proton collider with at least 10 TeV per beam.

Called the "Superconducting Super Collider" or SSC, this accelerator would have a circumference much larger than the 27 km of LEP. It is often nicknamed the "Desertron", as some people believe it would be too large to be accommodated elsewhere than in the desert of the American South-West. The SSC has barely been designed, let alone proposed officially, and it may face difficulty in being approved because of its high cost. Also, politicians may be sceptical after the recent cancellation of ISABELLE, a proton–proton collider formerly under construction at the Brookhaven National Laboratory. Nevertheless, Europeans are not letting the grass (sagebrush?) grow under their feet, and are already talking privately and quietly to each other about building a proton–proton or proton–anti-proton collider in the LEP tunnel, as a follow-up and complement to the e^+e^- collider already under construction.

There will surely be enough physics after the Z^0 to keep everybody on their toes, and a spirit of competition is always healthy. However, in view of the ever-increasing scale of high-energy particle accelerators, it seems ever more desirable that there should be the broadest possible international communication, consultation, co-operation and collaboration, with the aim of avoiding unnecessary duplication. Physicists have discussed for years the ideal of a "world accelerator" to which all interested countries would contribute. In today's world this seems to be a politically unrealisable dream. However, in the near future this may be the only basis on which experimental physicists can continue the dialogue with theorists that is essential for progress in fundamental physics.

17 May 1984

Index